Lecture Notes in Mechanics

Aristotle and Archytas defined mechanics as the *"organization of thought towards solving perplexing problems that are useful to humanity."* In the spirit Aristotle and Archytas, Lecture Notes in Mechanics (LNMech) transcends the traditional division of mechanics and provides a forum for presenting state-of-the-art research that tackles the range of complex issues facing society today.

LNMech provides for the rapid dissemination of comprehensive treatments of current developments in mechanics, serving as a repository and reference for innovation in mechanics, across all relevant application domains.

LNMech publishes original contributions, including monographs, extended surveys, and collected papers from workshops and conferences. All LNMech volumes are peer reviewed, available in print and online through ASCE, and indexed by the major scientific indexing and abstracting services.

LNMech 1 *Practical Approximate Analysis of Beams and Frames*
Nabil Fares, Ph.D.

LNMech 2 *Stochastic Models of Uncertainties in Computational Mechanics*
Christian Soize, Ph.D.

Stochastic Models of Uncertainties in Computational Mechanics

Christian Soize, Ph.D.

Library of Congress Cataloging-in-Publication Data on file.

Published by American Society of Civil Engineers
1801 Alexander Bell Drive
Reston, Virginia 20191
www.asce.org/pubs

ISBN 978-0-7844-1223-7 (paper)
ISBN 978-0-7844-7686-4 (e-book)

Manufactured in the United States of America.

18 17 16 15 14 13 12 1 2 3 4 5

Contents

Chapter 1

Introduction

This volume presents an overview of the main concepts, formulations, and recent advances in the area of stochastic modeling of uncertainties in computational mechanics. A mathematical-mechanical modeling process is useful for predicting the responses of a real structural system in its environment. Computational models are subject to two types of uncertainties—variabilities in the real system and uncertainties in the model itself—so, to be effective, these models must support robust optimization, design, and updating.

A probabilistic approach to uncertainties is the most powerful, efficient, and effective tool for computational modeling, as described in Chapter 2. The next three chapters focus on parametric (chapter 3), nonparametric (chapter 4), and generalized (chapter 5) probabilistic approaches to linear and nonlinear structural dynamics. Chapter 6 examines a nonparametric probabilistic approach to structural acoustics and vibration. Chapter 7 presents an extension of the nonparametric probabilistic approach to geometrically nonlinear elasticity in structural dynamics. Chapter 8 describes recent results in constructing prior probability models of non-Gaussian tensor-valued random fields based on the use of polynomial chaos expansion with applications to the mesoscale modeling of heterogeneous elastic microstructures.

Chapter 2

Short overview of probabilistic modeling of uncertainties and related topics

2.1 Uncertainty and variability

The *designed system* is used to manufacture the *real system* and to construct the nominal computational model (also called the *mean computational model* or sometimes, the mean model) using a mathematical-mechanical modeling process for which the main objective is the prediction of the responses of the real system in its environment. The real system, submitted to a given environment, can exhibit a variability in its responses due to fluctuations in the manufacturing process and due to small variations of the configuration around a nominal configuration associated with the designed system. The mean computational model which results from a mathematical-mechanical modeling process of the design system, has parameters which can be uncertain. In this case, there are *uncertainties on the computational model parameters*. In an other hand, the modeling process induces some modeling errors defined as the *model uncertainties*. Fig 2.1 summarizes the two types of uncertainties in a computational model and the variabilities of a real system. It is important to take into account both the uncertainties on the computational model parameters and the model uncertainties to improve the predictions of computational models in order to use such a computational model to carry out robust optimization, robust design and robust updating with respect to uncertainties. Today, it is well understood that, as soon as the probability theory can be used, then the probabilistic approach of uncertainties is certainly the most powerful, efficient and effective tool for

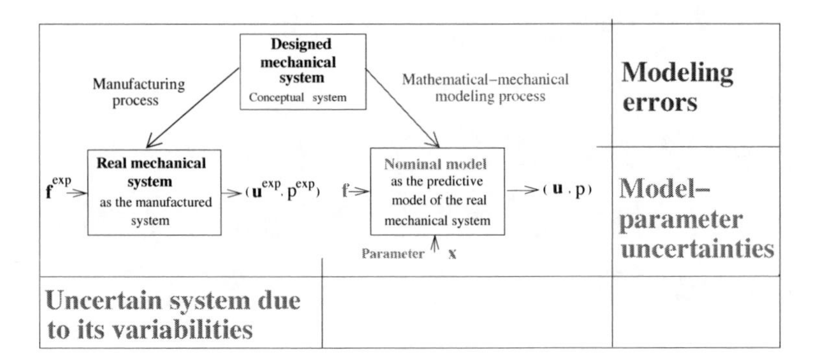

Figure 2.1. Variabilities and types of uncertainties in computational mechanics

modeling and for solving direct and inverse problem. The developments presented in this work are limited to the probabilistic approach.

2.2 Types of approach for probabilistic modeling of uncertainties

The *parametric probabilistic approach* consists in modeling the *uncertain parameters of the computational model* by random variables and then in constructing the probabilistic model of these random variables using the available information. Such an approach is very well adapted and very efficient to take into account the uncertainties on the computational model parameters as soon as the probability theory can be used. Many works have been published and a state-of-the-art can be found, for instance, in (Mace et al. 2005, Schueller 2005a;b; 2006; 2007, Deodatis and Spanos 2008, Schueller and Pradlwarter 2009a;b).

Concerning *model uncertainties* induced by *modeling errors*, it is today well understood that the prior and posterior probability models of the uncertain parameters of the computational model are not sufficient and do not have the capability to take into account model uncertainties in the context of computational mechanics as explained, for instance, in (Beck and Katafygiotis 1998) and in (Soize 2000; 2001; 2005b). Two main methods can be used to take into account model uncertainties (modeling errors).

(i) The first one consists in introducing a probabilistic model of the *output-prediction-error* which is the difference between the real system output and the computational model output. If there are no experimental data, then this method cannot really be used because there are generally no information concerning the probability model of the noise which is added to the computational model output. If experiments are available, the observed prediction error is then the difference between the measured real system output and the computational model output. A posterior probability model can then be constructed (see, for instance, (Beck and Katafygiotis 1998, Beck and Au 2002, Beck 2010)) using the Bayesian method (Bernardo and Smith 2000, Spall 2003, Kaipio and Somersalo 2005, Congdon 2007, Carlin and Louis 2009, Tan et al. 2010). Such an approach is efficient but requires experimental data. It should be noted that a lot of experimental data are required in high dimension. However, with such an approach, the posterior probability model of the uncertain parameters of the computational model strongly depends on the probability model of the noise which is added to the model output and which is often unknown. In addition, for many problems, it can be necessary to take into account the modeling errors at the operators level of the mean computational model. For instance, such an approach seems to be necessary to take into account the modeling errors on the mass and the stiffness operators of a computational dynamical model in order to analyze the generalized eigenvalue problem. It is also the case for the robust design optimization performed with an uncertain computational model for which the design parameters of the computational model are not fixed but run through an admissible set of values.

(ii) The second one is based on the *nonparametric probabilistic approach* of model uncertainties induced by modeling errors, which has been proposed in (Soize 2000) as an alternative method to the output-prediction-error method in order to take into account modeling errors at the operators level by introducing random operators and not at the model output level by introducing an additive noise. It should be noted that such an approach allows a prior probability model of model uncertainties to be constructed even if no experimental data are available. The nonparametric probabilistic approach is based on the use of a reduced-order model and the random matrix theory. It consists in directly constructing the stochastic modeling of the operators of the mean computational model. The random matrix theory (Mehta 1991) and its developments in the context of dynamics, vibration and acoustics (Soize 2000; 2001; 2005b; 2010c, Wright and Weaver 2010) is used to construct the prior probability distribution of the random matrices modeling the uncertain operators of the mean computational model. This prior probability distribution is constructed by using the Maximum Entropy Principle (Jaynes 1957), in the context of Information Theory (Shannon 1948), for which the constraints are defined by the available information (Soize 2000; 2001; 2003a; 2005a;b; 2010c). Since the paper (Soize 2000), many works have been published in order: (a) to validate the nonparametric probabilistic approach of both the computational-model-parameter uncertainties and the model uncertainties induced by modeling errors, with experimental results (Chebli and Soize 2004, Soize 2005b, Chen et al. 2006, Duchereau and Soize 2006, Soize et al. 2008a, Durand et al. 2008, Fernandez et al. 2009; 2010), (b) to extend the applicability of the theory to other areas (Soize 2003b, Soize and Chebli 2003, Capiez-Lernout and Soize 2004, Desceliers et al. 2004, Capiez-Lernout et al. 2005, Cottereau et al. 2007, Soize 2008a, Das and Ghanem 2009, Kassem et al. 2009), (c) to extend the theory to new ensembles of positive-definite random matrices yielding a more flexible description of the dispersion levels (Mignolet and Soize 2008a), (d) to apply the theory for the analysis of complex dynamical systems in the medium-frequency range, including structural-acoustic systems, (Ghanem and Sarkar 2003, Soize 2003b, Chebli and Soize 2004, Capiez-Lernout et al. 2006, Duchereau and Soize 2006, Arnst et al. 2006, Cottereau et al. 2008, Durand et al. 2008, Pellissetti et al. 2008, Desceliers et al. 2009, Fernandez et al. 2009; 2010, Kassem et al. 2011), and (e) to analyze nonlinear dynamical systems (i) with local nonlinear elements (Desceliers et al. 2004, Sampaio and Soize 2007a;b, Batou and Soize 2009b;a, Ritto et al. 2009; 2010, Wang et al. 2010) and (ii) with nonlinear geometrical effects (Mignolet and Soize 2007; 2008b).

Concerning the coupling of the parametric probabilistic approach of uncertain computational model parameters, with the nonparametric probabilistic approach of model uncertainties induced by modeling errors, a methodology has been proposed in (Soize 2010a, Batou et al. 2011). This *generalized probabilistic approach* of uncertainties in computational dynamics uses the random matrix theory. The proposed approach allows the prior probability model of each type of uncertainties (uncertainties on the computational model parameters and model uncertainties) to be separately constructed and identified. The modeling errors are not taken into account with the usual output-prediction-error method but with the nonparametric probabilistic approach.

2.3 Types of representation for the probabilistic modeling of uncertainties

A fundamental question is related to the construction of the prior probability model of uncertain computational model parameters but also to the construction of the prior probability model of model uncertainties induced by modeling errors. Such a prior probability model can then be used to study the propagation of uncertainties through the mechanical system which is analyzed. If experimental data are available for outputs of the mechanical system, then these experimental data can be used (1) to identify the parameters of the prior probability model using, for instance, the maximum likelihood method (Serfling 1980, Spall 2003) or (2) to construct a posterior probability model using the Bayesian method (Bernardo and Smith 2000, Spall 2003, Congdon 2007, Carlin and Louis 2009).

Two main methods are available to construct the prior probability model of a random vector $\theta \mapsto \mathbf{X}(\theta) = (X_1(\theta), \ldots, X_{n_p}(\theta))$ defined on a probability space $(\Theta, \mathcal{T}, \mathcal{P})$, with values in $\mathbb{R}^{n_p}$, and whose probability distribution on $\mathbb{R}^{n_p}$ is denoted by $P_{\mathbf{X}}(d\mathbf{x})$ (this random vector can be the approximation in finite dimension of a stochastic process or a random field). It should be noted that these two methods can also be used for random matrices, random fields, etc. We introduce the space $\mathcal{L}^2_{n_p}$ of all the second-order random variables $\mathbf{X}$ such that

$$E\{\|\mathbf{X}\|^2\} = \int_{\mathbb{R}^{n_p}} \|\mathbf{x}\|^2 \, P_{\mathbf{X}}(d\mathbf{x}) < +\infty,$$

in which E is the mathematical expectation, where $\mathbf{x} = (x_1, \ldots, x_{n_p})$ and $d\mathbf{x} = (dx_1 \ldots dx_{n_p})$, and where $\|\mathbf{x}\|^2 = x_1^2 + \ldots + x_{n_p}^2$ is the square of the Euclidean norm of vector $\mathbf{x}$.

(i)- The first method is a direct approach which consists in directly constructing the probability distribution $P_{\mathbf{X}}(d\mathbf{x})$ on $\mathbb{R}^{n_p}$ in using, for instance, the maximum entropy principle (see Section 2.4).

(ii)- The second one is an indirect approach which consists in introducing a representation $\mathbf{X} = \mathbf{h}(\boldsymbol{\Xi})$ for which $\mathbf{X}$ is the transformation by a deterministic nonlinear (measurable) mapping $\mathbf{h}$ (which has to be constructed) of a $\mathbb{R}^{N_g}$-valued random variable $\boldsymbol{\Xi} = (\Xi_1, \ldots, \Xi_{N_g})$ whose probability distribution $P_{\boldsymbol{\Xi}}(d\boldsymbol{\xi})$ is given and then is known. Then $P_{\mathbf{X}}$ is the transformation of $P_{\boldsymbol{\Xi}}$ by mapping $\mathbf{h}$. Two main types of methods can be used.

(ii.1) - The first one corresponds to the Polynomial Chaos representation (Ghanem and Spanos 1990; 1991; 2003) based on the Wiener mathematical works (Wiener 1938). The Polynomial Chaos representation of random variable $\mathbf{X}$ in $\mathcal{L}^2_{n_p}$ allows the effective construction of mapping $\mathbf{h}$ to be carried out as

$$\mathbf{X} = \Sigma_{j_1=0}^{+\infty} \ldots \Sigma_{j_{N_g}=0}^{+\infty} \, \mathbf{a}_{j_1,\ldots,j_{N_g}} \, \psi_{j_1}(\Xi_1) \times \ldots \times \psi_{j_{N_g}}(\Xi_{N_g}),$$

in which ψ_{j_k} are given real polynomials related to the probability distribution of $\boldsymbol{\Xi}$ and where $\mathbf{a}_{j_1,\ldots,j_{N_g}}$ are deterministic coefficients in $\mathbb{R}^{n_p}$.

Introducing the multi-index $\boldsymbol{\alpha} = (j_1, \ldots, j_{N_g})$ and the multi-dimensional polynomials $\psi_{\boldsymbol{\alpha}}(\boldsymbol{\Xi}) = \psi_{j_1}(\Xi_1) \times \ldots \times \psi_{j_{N_g}}(\Xi_{N_g})$ of random vector $\boldsymbol{\Xi}$, the previous polynomial chaos decomposition can be rewritten as

$$\mathbf{X} = \Sigma_{\boldsymbol{\alpha}}\, \mathbf{a}_{\boldsymbol{\alpha}}\, \psi_{\boldsymbol{\alpha}}(\boldsymbol{\Xi}).$$

The multi-dimensional polynomials $\psi_{\boldsymbol{\alpha}}(\boldsymbol{\Xi})$ are orthogonal and normalized,

$$E\{\psi_{\boldsymbol{\alpha}}(\boldsymbol{\Xi})\, \psi_{\boldsymbol{\beta}}(\boldsymbol{\Xi})\} = \int_{\mathbb{R}^{N_g}} \psi_{\boldsymbol{\alpha}}(\boldsymbol{\xi})\, \psi_{\boldsymbol{\beta}}(\boldsymbol{\xi})\, P_{\boldsymbol{\Xi}}(d\boldsymbol{\xi}) = \delta_{\boldsymbol{\alpha}\boldsymbol{\beta}},$$

in which $\delta_{\boldsymbol{\alpha}\boldsymbol{\alpha}} = 1$ and $\delta_{\boldsymbol{\alpha}\boldsymbol{\beta}} = 0$. The coefficients $\mathbf{a}_{\boldsymbol{\alpha}} = \mathbf{a}_{j_1,\ldots,j_{N_g}}$ are vectors in $\mathbb{R}^N$ which completely define mapping $\mathbf{h}$ and which are given by

$$\mathbf{a}_{\boldsymbol{\alpha}} = E\{\mathbf{X}\, \psi_{\boldsymbol{\alpha}}(\boldsymbol{\Xi})\}.$$

The construction of $\mathbf{h}$ then consists in identifying the vector-valued coefficients $\mathbf{a}_{\boldsymbol{\alpha}}$. If $\boldsymbol{\Xi}$ is a normalized Gaussian random vector, then the polynomials are the normalized Hermite polynomials.

The Polynomial Chaos representation can be applied in infinite dimension for stochastic processes and random fields (Ghanem and Kruger 1996, Sakamoto and Ghanem 2002, Xiu and Karniadakis 2002b, Debusschere et al. 2004, LeMaitre et al. 2004b;a, Lucor et al. 2004, Soize and Ghanem 2004, Wan and Karniadakis 2006, LeMaitre et al. 2007, Nouy 2007, Arnst and Ghanem 2008, Das et al. 2008, Ghosh and Ghanem 2008, Das et al. 2009, Nouy 2009, Soize and Ghanem 2009, LeMaitre and Knio 2010, Soize and Desceliers 2010).

Today, many applications using Polynomial Chaos representations have been carried out for modeling and solving direct and inverse problems in the domains of computational fluid mechanics and solid mechanics (Ghanem and Spanos 1990; 1991, Ghanem and Kruger 1996, Ghanem and Dham 1998, Ghanem and Red-Horse 1999, Ghanem 1999, Ghanem and Pellissetti 2002, LeMaitre et al. 2002, Xiu and Karniadakis 2002a; 2003, Ghanem et al. 2005, Wan and Karniadakis 2005, Ghanem and Doostan 2006, Berveiller et al. 2006, Desceliers et al. 2006, Faverjon and Ghanem 2006, Knio and LeMaitre 2006, Wan and Karniadakis 2006, Blatman and Sudret 2007, Ghanem and Ghosh 2007, Desceliers et al. 2007, LeMaitre et al. 2007, Lucor et al. 2007, Marzouk et al. 2007, Rupert and Miller 2007, Arnst and Ghanem 2008, Arnst et al. 2008, Ghanem et al. 2008, Guilleminot et al. 2008; 2009, Marzouk and Najm 2009, Najm 2009, Stefanou et al. 2009, LeMaitre and Knio 2010, Soize 2010b). We also refer the reader to Section 2.6 for a short overview concerning the analysis of the propagation of uncertainties and to Section 2.7 concerning the identification and inverse stochastic methods.

Sometimes, it can be interesting to replace the deterministic coefficients of the polynomial chaos decomposition by random coefficients. The theory of the polynomial chaos expansion with random coefficients of random vector $\mathbf{X}$ can be found in (Soize and Ghanem 2009) and is written as

$$\mathbf{X} = \Sigma_{\boldsymbol{\alpha}}\ \mathbf{A}_{\boldsymbol{\alpha}}\, \psi_{\boldsymbol{\alpha}}(\boldsymbol{\Xi}),$$

in which $\{\mathbf{A}_{\boldsymbol{\alpha}}\}_{\boldsymbol{\alpha}}$ are random coefficients with values in $\mathbb{R}^{n_p}$. Such a polynomial chaos representation with random coefficients is used in (Das et al. 2008, Ghanem et al. 2008, Soize and Ghanem 2009, Arnst et al. 2010). We will use such a representation in Section 8.6 to construct the posterior probability model of a random field.

(ii.2) - The second one consists in introducing a prior algebraic representation $\mathbf{X} = \mathbf{h}(\boldsymbol{\Xi}, \mathbf{s})$ in which $\mathbf{s}$ is a vector parameter which has a small dimension and which must be identified, where $\boldsymbol{\Xi}$ is a vector-valued random variable with a given probability distribution $P_{\boldsymbol{\Xi}}$ and where $\mathbf{h}$ is a given nonlinear mapping. For instance, tensor-valued random fields representations constructed with such an approach can be found in (Soize 2006; 2008b, Guilleminot et al. 2011, Guilleminot and Soize 2011). We will present an example in Section 8.

In theory, method (ii.1) allows any random vector $\mathbf{X}$ in $\mathcal{L}^2_{n_p}$ to be represented by a Polynomial Chaos expansion. In practice, the representation can require a very large number of coefficients to get convergence yielding very difficult problems in high dimension for the identification problem and then requires adapted methodologies and adapted methods (Soize and Desceliers 2010, Soize 2010b; 2011). In general, method (ii.2) does not allow any random vector $\mathbf{X}$ in $\mathcal{L}^2_{n_p}$ to be represented but allows a family of representations to be constructed in a subset of $\mathcal{L}^2_{n_p}$ when $\mathbf{s}$ runs through its admissible set (but, in counter part, the identification of $\mathbf{s}$ is effective and efficient).

2.4 Construction of prior probability models using the maximum entropy principle under the constraints defined by the available information

The measure of uncertainties using the entropy has been introduced by (Shannon 1948) in the framework of Information Theory. The maximum entropy principle (that is to say the maximization of the level of uncertainties) has been introduced by (Jaynes 1957) to construct the prior probability model of a random variable under the constraints defined by the available information. This principle appears as a major tool to construct the prior probability model (1) of the uncertain parameters of the computational model using the parametric probabilistic approach, (2) of both the computational-model-parameter uncertainties and the model uncertainties, using the nonparametric probabilistic approach and (3) of the generalized approach of uncertainties corresponding to a full coupling of the parametric and the nonparametric probabilistic approaches.

Let $\mathbf{x} = (x_1, \ldots, x_{n_p})$ be a real vector and let $\mathbf{X} = (X_1, \ldots, X_{n_p})$ be a second-order random variable with values in $\mathbb{R}^{n_p}$ whose probability distribution $P_{\mathbf{X}}$ is defined by a probability density function $\mathbf{x} \mapsto p_{\mathbf{X}}(\mathbf{x})$ on $\mathbb{R}^{n_p}$ with respect to $d\mathbf{x} = dx_1 \ldots dx_{n_p}$ and which verifies the normalization condition $\int_{\mathbb{R}^{n_p}} p_{\mathbf{X}}(\mathbf{x})\, d\mathbf{x} = 1$. It is assumed that $\mathbf{X}$ is with values in any bounded or unbounded part $\mathcal{X}$ of $\mathbb{R}^{n_p}$ and consequently, the support of $p_{\mathbf{X}}$ is $\mathcal{X}$. The available information defines a constraint equation on $\mathbb{R}^{\mu}$ written as

$$E\{\mathbf{g}(\mathbf{X})\} = \mathbf{f},$$

in which E is the mathematical expectation, $\mathbf{f}$ is a given vector in $\mathbb{R}^{\mu}$ and where $\mathbf{x} \mapsto \mathbf{g}(\mathbf{x})$ is a given function (measurable) from $\mathbb{R}^{n_p}$ into $\mathbb{R}^{\mu}$. Since $p_{\mathbf{X}}$ is unknown and has to be constructed, we introduce the set $\mathcal{C}$ of all the probability density functions $\mathbf{x} \mapsto p(\mathbf{x})$ defined on $\mathbb{R}^{n_p}$, with values in $\mathbb{R}^{+}$, with support $\mathcal{X}$, verifying the normalization condition and the constraint equation $E\{\mathbf{g}(\mathbf{X})\} = \mathbf{f}$, that is to say,

$$\mathcal{C} = \left\{ \mathbf{x} \mapsto p(\mathbf{x}) \geq 0 \,,\ \text{Supp}\, p = \mathcal{X} \,, \int_{\mathcal{X}} p(\mathbf{x})\, d\mathbf{x} = 1 \,, \int_{\mathcal{X}} \mathbf{g}(\mathbf{x})\, p(\mathbf{x})\, d\mathbf{x} = \mathbf{f} \right\}.$$

The maximum entropy principle consists in finding $p_{\mathbf{X}}$ in $\mathcal{C}$ which maximizes the entropy (that is to say which maximizes the uncertainties),

$$p_{\mathbf{X}} = \arg \max_{p \in \mathcal{C}} \, S(p), \quad S(p) = -\int_{\mathbb{R}^N} p(\mathbf{x}) \, \log(p(\mathbf{x}))\, d\mathbf{x},$$

in which $S(p)$ is the entropy of the probability density function p. Introducing the Lagrange multiplier $\boldsymbol{\lambda} \in \mathcal{L}_{\mu} \subset \mathbb{R}^{\mu}$ (associated with the constraint), where $\mathcal{L}_{\mu}$ is the subset of $\mathbb{R}^{\mu}$ of all the admissible values for $\boldsymbol{\lambda}$, it can easily be seen that the solution of the optimization problem can be written as

$$p_{\mathbf{X}}(\mathbf{x}) = c_0 \, \mathbb{1}_{\mathcal{X}}(\mathbf{x}) \, \exp(- < \boldsymbol{\lambda}, \mathbf{g}(\mathbf{x}) >), \quad \forall \mathbf{x} \in \mathbb{R}^{n_p}$$

in which $< \mathbf{x}\,, \mathbf{y} >= x_1 y_1 + \ldots + x_{\mu} y_{\mu}$ and where $\mathbb{1}_{\mathcal{X}}$ is the indicator function of set $\mathcal{X}$. The normalization constant c_0 and the Lagrange multiplier

$\boldsymbol{\lambda}$ are calculated in solving the following nonlinear vectorial algebraic equations

$$c_0 \int_{\mathcal{X}} \exp(- < \boldsymbol{\lambda}, \mathbf{g}(\mathbf{x}) >) \, d\mathbf{x} = 1,$$

$$c_0 \int_{\mathcal{X}} \mathbf{g}(\mathbf{x}) \exp(- < \boldsymbol{\lambda}, \mathbf{g}(\mathbf{x}) >) \, d\mathbf{x} = \mathbf{f}.$$

These algebraic equations can be solved using appropriated algorithms. Then, it is necessary to construct a generator of independent realizations of random variable $\mathbf{X}$ whose probability density function is the one which has been built. In small dimension (n_p is a few units), there is no difficulty. In high dimension (n_p several hundreds or several thousands), there are two major difficulties. The first one is related to the calculation of an integral in high dimension of the type $c_0 \int_{\mathcal{X}} \mathbf{g}(\mathbf{x}) \exp(- < \boldsymbol{\lambda}, \mathbf{g}(\mathbf{x}) >) \, d\mathbf{x}$. Such a calculation is necessary to implement the algorithm for computing c_0 and $\boldsymbol{\lambda}$. The second one is the construction of the generator once c_0 et $\boldsymbol{\lambda}$ have been calculated. These two aspects can be solved using the Markov Chain Monte Carlo methods (MCMC) (MacKeown 1997, Spall 2003, Kaipio and Somersalo 2005, Rubinstein and Kroese 2008). The transition kernel of the homogeneous (stationary) Markov chain of the MCMC method can be constructed using the Metropolis-Hastings algorithm (Metropolis and Ulam 1949, Metropolis et al. 1953, Hastings 1970, Casella and George 1992) or the Gibbs algorithm (Geman and Geman 1984) which is a slightly different algorithm for which the kernel is directly derived from the transition probability density function and for which the Gibbs realizations are always accepted. These two algorithms construct the transition kernel for which the invariant measure is $P_{\mathbf{X}}$. In general, these algorithms are effective but can not be when there are attraction regions which do not correspond to the invariant measure. These situations can not be easily detected and are time consuming. An alternative approach (Soize 2008a) of the class of the Gibbs method has been proposed to avoid these difficulties. This approach is based on the introduction of an Itô stochastic differential equation whose unique invariant measure is $P_{\mathbf{X}}$. This invariant measure is then the explicit solution of a Fokker-Planck equation (Soize 1994). The algorithm is then obtained by the discretization of the Itô equation.

2.5 Random Matrix Theory

The random matrix theory were introduced and developed in mathematical statistics by Wishart and others in the 1930s and was intensively studied by physicists and mathematicians in the context of nuclear physics. These works began with Wigner in the 1950s and received an important effort in the 1960s by Wigner, Dyson, Mehta and others. In 1965, Poter published a volume of important papers in this field, followed, in 1967 by the first edition of the Mehta book whose second edition (Mehta 1991) published in 1991 is an excellent synthesis of the random matrix theory. We refer the reader to (Wright and Weaver 2010) and (Soize 2010c) for an introduction to random matrix theory presented in the context of Mechanics. Concerning multivariate statistical analysis and statistics of random matrices, the reader will find additional developments in Fougeaud and Fuchs (Fougeaud and Fuchs 1967) and Anderson (Anderson 1958).

2.5.1 Why the Gaussian orthogonal ensemble cannot be used if positiveness property is required

The Gaussian Orthogonal Ensemble (GOE), for which the mean value is the unity matrix $[I_n]$ (Soize 2003a), is the set of the random matrices $[\mathbf{G}^{\rm GOE}]$) which can be written as $[\mathbf{G}^{\rm GOE}] = [I_n] + [\mathbf{B}^{\rm GOE}]$ in which $[\mathbf{B}^{\rm GOE}]$ belongs to the GOE (Mehta 1991), for which the mean value is the zero matrix, $[0]$, and consequently, would be a second-order centered random matrix with values in $\mathbb{M}_n^S(\mathbb{R})$ such that $E\{[\mathbf{B}^{\rm GOE}]\} = [\,0\,]$ and $E\{\|[\mathbf{B}^{\rm GOE}]\|_F^2\} < +\infty$, in which $\|A\|_F^2 = \text{tr}\{[A]^T[A]\} = \sum_{j=1}^n \sum_{k=1}^n [A]_{jk}^2$ is the square of the Frobenius norm of the symmetric real matrix $[A]$. The probability density function of random matrix $[\mathbf{B}^{\rm GOE}]$, with respect to $\widetilde{d}[B] = 2^{n(n-1)/4}\,\Pi_{1\leq i\leq j\leq n}\, d[B]_{ij}$, is written as

$$p_{[\mathbf{B}^{\rm GOE}]}([B]) = C_n \times \exp\left\{-\frac{(n+1)}{4\delta^2}\text{tr}\{[B]^2\}\right\}.$$

The constant C_n of normalization can easily be calculated and δ is the coefficient of variation of the random matrix $[\mathbf{G}^{\rm GOE}]$ which is such that $\delta^2 = n^{-1}E\{\|\,[\mathbf{G}^{\rm GOE}] - [I_n]\,\|_F^2\}$ because $\|\,[I_n]\,\|_F^2 = n$. The real-valued random variables $\{[\mathbf{B}^{\rm GOE}]_{jk}, j \leq k\}$ are statistically independent, second order, centered and Gaussian. It can be seen that $[\mathbf{G}^{\rm GOE}]$ is with values in $\mathbb{M}_n^S(\mathbb{R})$ but is not positive definite. In addition,

$$E\{\|[\mathbf{G}^{\rm GOE}]^{-1}\|^2\} = +\infty.$$

Consequently, $[\mathbf{G}^{\rm GOE}]$ is not acceptable if positiveness property and integrability of the inverse are required.

2.5.2 Ensemble SG_0^+ of random matrices

The GOE cannot be used when positiveness property and integrability of the inverse are required. Consequently, we need new ensembles of random matrices which will be used to develop the nonparametric probabilistic approach of uncertainties in computational solid and fluid mechanics, and which differ from the GOE and from the other known ensembles of the random matrix theory.

The objective of this section is then to summarize the construction given in (Soize 2000; 2001; 2003a; 2005b) of the ensemble SG_0^+ of random matrices $[\mathbf{G}_0]$ defined on the probability space $(\Theta, \mathcal{T}, \mathcal{P})$, with values in the set $\mathbb{M}_n^+(\mathbb{R})$ of all the positive definite symmetric $(n \times n)$ real matrices and such that

$$E\{[\mathbf{G}_0]\} = [I_n], \quad E\{\log(\det[\mathbf{G}_0])\} = C, \quad |C| < +\infty.$$

The probability distribution $P_{[\mathbf{G}_0]} = p_{[\mathbf{G}_0]}([G])\, \widetilde{dG}$ is defined by a probability density function $[G] \mapsto p_{[\mathbf{G}_0]}([G])$ from $\mathbb{M}_n^+(\mathbb{R})$ into $\mathbb{R}^+$ with respect to the volume element $\widetilde{dG}$ on the set $\mathbb{M}_n^S(\mathbb{R})$ of all the symmetric $(n \times n)$ real matrices, which is such that $\widetilde{dG} = 2^{n(n-1)/4} \Pi_{1 \leq j \leq k \leq n} dG_{jk}$. This probability density function can then verify the normalization condition,

$$\int_{\mathbb{M}_n^+(\mathbb{R})} p_{[\mathbf{G}_0]}([G])\, \widetilde{dG} = 1.$$

Let δ be the positive real number defined by

$$\delta = \left\{ \frac{E\{\| [\mathbf{G}_0] - E\{[\mathbf{G}_0]\} \|_F^2\}}{\| E\{[\mathbf{G}_0]\} \|_F^2} \right\}^{1/2} = \left\{ \frac{1}{n} E\{\| [\mathbf{G}_0] - [I_n] \|_F^2\} \right\}^{1/2},$$

which will allow the dispersion of the probability model of random matrix $[\mathbf{G}_0]$ to be controlled. For δ such that $0 < \delta < (n+1)^{1/2}(n+5)^{-1/2}$, the use of the maximum entropy principle under the constraints defined by the above available information yields the following algebraic expression of the probability density function of random matrix $[\mathbf{G}_0]$,

$$p_{[\mathbf{G}_0]}([G]) = \mathbb{1}_{\mathbb{M}_n^+(\mathbb{R})}([G]) \times C_{\mathbf{G}_0} \times \left(\det[G]\right)^{(n+1)\frac{(1-\delta^2)}{2\delta^2}} \times e^{-\frac{(n+1)}{2\delta^2}\mathrm{tr}[G]},$$

in which $\mathbb{1}_{\mathbb{M}_n^+(\mathbb{R})}([G])$ is equal to 1 if $[G] \in \mathbb{M}_n^+(\mathbb{R})$ and is equal to zero if $[G] \notin \mathbb{M}_n^+(\mathbb{R})$, where $\mathrm{tr}[G]$ is the trace of matrix $[G]$, where $\det[G]$ is the determinant of matrix $[G]$ and where the positive constant $C_{\mathbf{G}_0}$ is such that

$$C_{\mathbf{G}_0} = (2\pi)^{-n(n-1)/4} \left(\frac{n+1}{2\delta^2}\right)^{n(n+1)(2\delta^2)^{-1}} \left\{ \Pi_{j=1}^n \Gamma\left(\frac{n+1}{2\delta^2} + \frac{1-j}{2}\right) \right\}^{-1},$$

and where, for all $z > 0$, $\Gamma(z) = \int_0^{+\infty} t^{z-1} e^{-t}\, dt$. Note that $\{[\mathbf{G}_0]_{jk}, 1 \leq j \leq k \leq n\}$ are dependent random variables. If $(n+1)/\delta^2$ is an integer, then this probability density function coincides with the Wishart probability distribution (Anderson 1958, Fougeaud and Fuchs 1967). If $(n+1)/\delta^2$ is not an integer, then this probability density function can be viewed as a particular case of the Wishart distribution, in infinite dimension, for stochastic processes (Soize 1980).

Let $\|G\| = \sup_{\|\mathbf{x}\|\leq 1} \|[G]\,\mathbf{x}\|$ be the operator norm of matrix $[G]$ which is such that $\|[G]\,\mathbf{x}\| \leq \|G\|\,\|\mathbf{x}\|$ for all $\mathbf{x}$ in $\mathbb{R}^n$. Let $\|G\|_F$ be the Frobenius norm of $[G]$ which is defined by $\|G\|_F^2 = \text{tr}\{[G]^T[G]\} = \sum_{j=1}^n \sum_{k=1}^n [G]_{jk}^2$ and which is such that $\|G\| \leq \|G\|_F \leq \sqrt{n}\,\|G\|$. It is proven (Soize 2001) that

$$E\{\|[\mathbf{G}_0]^{-1}\|^2\} \leq E\{\|[\mathbf{G}_0]^{-1}\|_F^2\} < +\infty.$$

In general, the above equation does not imply that $n \mapsto E\{\|[\mathbf{G}_0]^{-1}\|^2\}$ is a bounded function with respect to n, but, in the present case, we have the following fundamental property,

$$\forall n \geq 2, \quad E\{\|[\mathbf{G}_0]^{-1}\|^2\} \leq C_\delta < +\infty,$$

in which C_δ is a positive finite constant which is independent of n but which depends on δ. The above equation means that $n \mapsto E\{\|[\mathbf{G}_0]^{-1}\|^2\}$ is a bounded function from $\{n \geq 2\}$ into $\mathbb{R}^+$.

The generator of independent realizations (which is required to solve the random equations with the Monte Carlo method) can be constructed (Soize 2000; 2001) using the following algebraic representation. Random matrix $[\mathbf{G}_0]$ can be written (Cholesky decomposition) as $[\mathbf{G}_0] = [\mathbf{L}]^T\,[\mathbf{L}]$ in which $[\mathbf{L}]$ is an upper triangular $(n \times n)$ random matrix such that:

(1) random variables $\{[\mathbf{L}]_{jj'}, j \leq j'\}$ are independent;

(2) for $j < j'$, the real-valued random variable $[\mathbf{L}]_{jj'}$ is written as $[\mathbf{L}]_{jj'} = \sigma_n U_{jj'}$ in which $\sigma_n = \delta(n+1)^{-1/2}$ and where $U_{jj'}$ is a real-valued Gaussian random variable with zero mean and variance equal to 1;

(3) for $j = j'$, the positive-valued random variable $[\mathbf{L}]_{jj}$ is written as $[\mathbf{L}]_{jj} = \sigma_n \sqrt{2V_j}$ in which σ_n is defined above and where V_j is a positive-valued gamma random variable whose probability density function is $p_{V_j}(v) = \mathbb{1}_{\mathbb{R}^+}(v)\frac{1}{\Gamma(a_j)}\, v^{a_j-1}\, e^{-v}$, in which $a_j = \frac{n+1}{2\delta^2} + \frac{1-j}{2}$. It should be noted that the probability density function of each diagonal element $[\mathbf{L}]_{jj}$ of the random matrix $[\mathbf{L}]$ depends on the rank j of the element.

2.5.3 Ensemble SG_ε^+ of random matrices

Let $\mathcal{L}_n^2$ be the set of all the second-order random variables, defined on the probability space $(\Theta, \mathcal{T}, P)$, with values in $\mathbb{R}^n$, equipped with the inner product $\ll \mathbf{X}, \mathbf{Y} \gg = E\{< \mathbf{X}, \mathbf{Y} >\}$ and with the associated norm $|||\mathbf{X}||| = \ll \mathbf{X}, \mathbf{X} \gg^{1/2}$. Let $0 \leq \varepsilon \ll 1$ be a positive number as small as one wants. The ensemble $\mathrm{SG}_\varepsilon^+$ is defined as the ensemble of all the random matrices such that

$$[\mathbf{G}] = \frac{1}{1+\varepsilon}\{[\mathbf{G}_0] + \varepsilon\,[I_n]\},$$

in which $[\mathbf{G}_0]$ is a random matrix which belongs to ensemble SG_0^+. Let $[\mathbf{G}]$ be in $\mathrm{SG}_\varepsilon^+$ with $\varepsilon \geq 0$ fixed as small as one wants (possibly, ε can be equal to zero and in such a case, $\mathrm{SG}_\varepsilon^+ = \mathrm{SG}_0^+$ and then, $[\mathbf{G}] = [\mathbf{G}_0]$). It can easily be seen that

$$E\{[\mathbf{G}]\} = [I_n].$$

Let $\mathbb{b}(\mathbf{X}, \mathbf{Y})$ be the bilinear form on $\mathcal{L}_n^2 \times \mathcal{L}_n^2$ such that $\mathbb{b}(\mathbf{X}, \mathbf{Y}) = \ll [\mathbf{G}]\,\mathbf{X}, \mathbf{Y} \gg$. For all $\mathbf{X}$ in $\mathcal{L}_n^2$, we have

$$\mathbb{b}(\mathbf{X}, \mathbf{X}) \geq c_\varepsilon |||\mathbf{X}|||^2$$

in which $c_\varepsilon = \varepsilon/(1+\varepsilon)$. The proof can easily be obtained. We have $\mathbb{b}(\mathbf{X}, \mathbf{X}) = 1/(1+\varepsilon) \ll [\mathbf{G}_0]\,\mathbf{X}, \mathbf{X} \gg + \varepsilon/(1+\varepsilon)|||\mathbf{X}|||^2 \geq c_\varepsilon|||\mathbf{X}|||^2$, because, for all $[\mathbf{G}_0]$ in SG_0^+, and for all $\mathbf{X}$ in $\mathcal{L}_n^2$, we have $\ll [\mathbf{G}_0]\,\mathbf{X}, \mathbf{X} \gg \geq 0$. Finally, for all $\varepsilon \geq 0$, it can be proven that

$$E\{\|[\mathbf{G}]^{-1}\|_F^2\} < +\infty.$$

2.6 Propagation of uncertainties and methods to solve the stochastic dynamical equations

Concerning the methods and formulations to analyze the propagation of uncertainties in the computational model, the choice of a specific method depends on the desired accuracy on the model output and on the nature of the expected probabilistic information. These last two decades, a growing interest has been devoted to spectral stochastic methods, pioneered by Roger Ghanem in 1990-1991 (Ghanem and Spanos 1990; 1991; 2003), and which provide an explicit representation of the random model output as a function of the basic random parameters modeling the input uncertainties (Ghanem and Spanos 1991, Ghanem and Kruger 1996, Ghanem and Red-Horse 1999, Ghanem 1999, Xiu and Karniadakis 2002b, Ghanem and Spanos 2003, Ghosh and Ghanem 2008, Matthies 2008, Nouy 2009, LeMaitre and Knio 2010). An approximation of the random model output is sought on suitable functional approximation bases.

There are two distinct classes of techniques for the definition and computation of this approximation.

(1) The first class of techniques rely on Galerkin-type projections of weak solutions of models involving differential or partial differential equations (Ghanem and Spanos 1991; 2003, Deb et al. 2001, LeMaitre et al. 2004b;a, Babuska et al. 2005, Frauenfelder et al. 2005, Matthies and Keese 2005, Wan and Karniadakis 2005; 2006, Mathelin and LeMaitre 2007, Wan and Karniadakis 2009). Alternative methods, based on the construction of optimal separated representations of the solution, have been performed to reduce computational requirements. They consist in representing the solution on optimal reduced bases of deterministic functions and stochastic functions (scalar-valued random variables). Several strategies are proposed for the construction of reduced bases using approximate Karhunen-Loeve expansions (or classical spectral decompositions) of the solution (Matthies and Keese 2005, Doostan et al. 2007). Another method which is called Generalized Spectral Decomposition method, can be used to construct such representations without knowing *a priori* the solution nor an approximation of it (Nouy 2007; 2008, Nouy and LeMaitre 2009). An advantage of these algorithms is that they allow a separation of deterministic problems for the computation of deterministic functions and stochastic algebraic equations for the computation of stochastic functions. In that sense, they can be considered as partly non-intrusive techniques for computing Galerkin-type projections. Similarly, multidimensional extensions of separated representations techniques have been proposed (Doostan and Iaccarino 2009, Nouy 2010). These methods exploit the tensor product structure of the solution function space, resulting from the product structure of the probability space defined by input random parameters. However, it does not circumvent the curse of dimensionality associated with the dramatic increase in the dimension of stochastic approximation spaces when dealing with high stochastic dimension and high approximation resolution along each stochastic dimension.

(2) The second class is composed of methods based on a direct simulation (Matthies and Keese 2005, Berveiller et al. 2006, Babuska et al. 2007, Blatman and Sudret 2007, Ganapathysubramanian and Zabaras 2007, Webster et al. 2007, Zabaras and Ganapathysubramanian 2008, Ma and Zabaras 2009). These methods are often called non-intrusive since they offer the advantage of only requiring the availability of classical deterministic codes. Finally, the direct Monte Carlo numerical simulation method (see for instance (Fishman 1996, MacKeown 1997, Rubinstein and Kroese 2008) is a very effective and efficient method because this

method (i) is non-intrusive, (ii) is adapted to massively parallel computation without any software developments, (iii) is such that its convergence can be controlled during the computation, and (iv) the speed of convergence is independent of the dimension. The speed of convergence of the Monte Carlo method can be improved using advanced Monte Carlo simulation procedures (Papadrakakis and Papadopoulos 1996, Pradlwarter and Schueller 1997, Papadrakakis and Kotsopulos 1999, Schueller 2009), subset simulation technics (Au and Beck 2003b), important sampling for high dimension problems (Au and Beck 2003a), local domain Monte Carlo Simulation (Pradlwarter and Schueller 2010).

2.7 Identification of the prior and posterior probability models of uncertainties

The identification of the parameters of the probability model of uncertainties (parametric and nonparametric probabilistic approaches) is a problem belonging to the class of the statistical inverse problems, see for instance (Kaipio and Somersalo 2005).

2.7.1 Identification of the probability model of random variables in the context of the parametric probabilistic approach of model-parameter uncertainties

Let us assume that experimental data are available for observations related to the random computational model output. The experimental identification of the parameters of the prior probability distributions of the random variables which model the uncertain parameters in the computational model can be performed, either in minimizing a distance between the experimental observations and the random model observations (such as least-square method), or in using the maximum likelihood method (Serfling 1980, Spall 2003). Such an approach is described in Sections 3.5, 5.3 and 8.3, and many applications can be found in the literature. In the domain of structural dynamics, vibrations and structural acoustics, we refer the reader to (Peters and Roeck 2001, Ghanem et al. 2005, Arnst et al. 2008, Goller et al. 2009) for the stochastic system identification in structural dynamics and vibration, to (Faverjon and Ghanem 2006) for the stochastic inversion in acoustic scattering, to (Cataldo et al. 2009) for the probabilistic modeling of a nonlinear dynamical system used for producing voice and to (Leissing et al. 2010) for the computational model for long-range non-linear propagation over urban cities. With such an identification, we then obtained an optimal prior probability distribution. The Bayesian method then allows a posterior probability distribution to be constructed from the optimal prior probability distribution and from the experimental data. Many works have been published in the literature: see, for instance, the textbooks on the Bayesian method such as (Bernardo and Smith 2000, Kaipio and Somersalo 2005, Congdon 2007, Carlin and Louis 2009) and papers devoted to the use of the Bayesian method in the context of uncertain mechanical and dynamical systems such as(Beck and Katafygiotis 1998, Katafygiotis and Beck 1998, Beck and Au 2002, Ching et al. 2006, Cheung and Beck 2009, Beck 2010, Cheung and Beck 2010, Tan et al. 2010). We will use such a Bayesian approach in Sections 3.6, 5.4 and 8.6.

2.7.2 Identification of the probability model of random matrices in the context of the nonparametric probabilistic approach of both model-parameter uncertainties and model uncertainties

General developments concerning the experimental identification of the parameters of the prior probability distributions of random matrices used to model the uncertainties in computational mechanics can be found in (Soize 2005b). Many works have been published in this field, such as (Chebli and Soize 2004, Arnst et al. 2006; 2008, Chen et al. 2006, Duchereau and Soize 2006, Soize et al. 2008a, Batou et al. 2011) for the experimental identification of the nonparametric probabilistic model of uncertainties in structural dynamics, as (Durand et al. 2008, Fernandez et al. 2010) in structural acoustics for the low- and medium-frequency ranges and as (Batou and Soize 2009b;a) for nonlinear structural dynamical systems. The identification of the generalized probabilistic approach of uncertainties can be found in (Soize 2010a, Batou et al. 2011) and also in Section 5.3.

2.7.3 Identification of the probability model of random fields in the context of the parametric probabilistic approach of model-parameter uncertainties

Concerning the identification of prior probability models of random fields for which the parametric stochastic modeling is parameterized with a small number of parameters, efficient methods have been developed based on least-square approach, maximum likelihood method, etc, and have been applied with success to several areas for linear and nonlinear problems (Arnst et al. 2008, Batou and Soize 2009b, Guilleminot et al. 2008, Batou and Soize 2009a). The stochastic inverse methods and the Bayesian inference approach to inverse problems have received a particular attention (Faverjon and Ghanem 2006, Ma and Zabaras 2009, Marzouk et al. 2007, Marzouk and Najm 2009, Wang and Zabaras 2004; 2005, Zabaras and Ganapathysubramanian 2008, Arnst et al. 2010). The problem relative to the identification with experimental data of the deterministic coefficients of the polynomial chaos expansion in small dimension of a non-Gaussian real-valued random field using the maximum likelihood has been introduced in (Desceliers et al. 2006; 2007) and revisited in (Das et al. 2009). In practice, the identification of these deterministic coefficients is performed using a finite length of experimental data sets. Consequently, the maximum likelihood statistical estimator of these coefficients is not completely converged and therefore, there are residual statistical fluctuations which can be modeled in representing the deterministic coefficients by random coefficients. In (Das et al. 2008), it is proposed to construct the probability model of these random coefficients by using the asymptotic sampling Gaussian distribution constructed with the Fisher information matrix and which is a consistent and asymptotically efficient estimator. Such an approach has been used for model validation (Ghanem et al. 2008, Red-Horse and Benjamin 2004). In (Arnst et al. 2010), as a continuation of (Soize and Ghanem 2009), the identification of Bayesian posteriors for the coefficients of chaos expansions is proposed: the usual deterministic coefficients of the chaos expansion are replaced by random coefficients using the formalism introduced in (Soize and Ghanem 2009) in order to quantify the uncertainties induced by the errors introduced by the use of a relatively low order of the maximum degree of polynomials in the chaos expansion and induced by the errors due to the finite length of the experimental data set. These methodologies is efficient in low dimension but cannot easily be extended in high dimension. Nevertheless, it constitutes a solid basis to start an extension for high dimension. In this context, a methodology for the identification of high-dimension polynomial chaos expansions with random coefficients for non-Gaussian tensor-valued random fields using partial and limited data is proposed in (Soize 2010b, Soize and Desceliers 2010, Soize 2011). Such an approach is developed in Section 8.

2.8 Robust updating of computational models and robust design with uncertain computational models

Robust updating or robust design optimization consists in updating a computational model or in optimizing the design of a mechanical system with a computational model, in taking into account both the uncertainties in the computational model parameters and the model uncertainties. An overview on computational methods in optimization considering uncertainties can be found in (Schueller and Jensen 2008). Robust updating and robust design developments with uncertainties on the computational model parameters are developed in (Beck et al. 1999, Papadimitriou et al. 2000; 2001, Papadrakakis and Lagaros 2002, Taflanidis and Beck 2008, Goller et al. 2009) while robust updating or robust design optimization with model uncertainties can be found in (Capiez-Lernout and Soize 2008b;a;c, Soize et al. 2008b, Ritto et al. 2010).

Chapter 3

Parametric probabilistic approach to uncertainties in computational structural dynamics

3.1 Introduction of the mean computational model in computational structural dynamics

The developments are presented for computational models in structural dynamics. The dynamical system is a damped fixed structure around a static equilibrium configuration considered as a natural state without prestresses and subjected to an external load. For given nominal values of the parameters of the dynamical system, the basic finite element model is called the *mean computational model*. In addition, it is assumed that a set of model parameters has been identified as the uncertain model parameters. These model parameters are the components of a vector $\mathbf{x} = (x_1, \ldots, x_{n_p})$ belonging to an admissible set $\mathcal{C}_{\text{par}}$ which is a subset of $\mathbb{R}^{n_p}$. Using the usual finite element method to construct a finite approximation of the weak formulation of the boundary value problem yields the dynamical equation of the mean computational model which is then written as

$$[\,\mathbb{M}(\mathbf{x})]\,\ddot{\mathbb{y}}(t) + [\,\mathbb{D}(\mathbf{x})]\,\dot{\mathbb{y}}(t) + [\,\mathbb{K}(\mathbf{x})]\,\mathbb{y}(t) + \mathbb{f}_{\text{NL}}(\mathbb{y}(t), \dot{\mathbb{y}}(t); \mathbf{x}) = \mathbb{f}(t; \mathbf{x}), \quad (3.1)$$

in which $\mathbb{y}(t)$ is the unknown time response vector of the m degrees of freedom (DOF) (displacements and/or rotations); $\dot{\mathbb{y}}(t)$ and $\ddot{\mathbb{y}}(t)$ are the velocity and acceleration vectors respectively; $\mathbb{f}(t; \mathbf{x})$ is the known external load vector of the m inputs (forces and/or moments); $[\,\mathbb{M}(\mathbf{x})]$, $[\,\mathbb{D}(\mathbf{x})]$

and $[\mathbb{K}(\mathbf{x})]$ are the mass, damping and stiffness matrices of the mean computational model, which belong to the set $\mathbb{M}_m^+(\mathbb{R})$ of all the positive-definite symmetric $(m \times m)$ real matrices; $(\mathbf{y}, \mathbf{z}) \mapsto \mathbb{f}_{\text{NL}}(\mathbf{y}, \mathbf{z}; \mathbf{x})$ is the nonlinear mapping modeling local nonlinear forces.

3.2 Introduction of the reduced mean computational model

In the context of the parametric probabilistic approach of model-parameter uncertainties, parameter $\mathbf{x}$ is modeled by a random variable $\mathbf{X}$. The mean value of $\mathbf{X}$ is chosen as the nominal value $\underline{\mathbf{x}} = (\underline{x}_1, \ldots, \underline{x}_{n_p})$ of the uncertain model parameter $\mathbf{x}$. The support of the probability distribution of $\mathbf{X}$ on $\mathbb{R}^{n_p}$ is $\mathcal{C}_{\text{par}} \subset \mathbb{R}^{n_p}$.

For all $\mathbf{x}$ fixed in $\mathcal{C}_{\text{par}}$, let $\{\boldsymbol{\phi}_1(\mathbf{x}), \ldots, \boldsymbol{\phi}_m(\mathbf{x})\}$ be an algebraic basis of $\mathbb{R}^m$ constructed, for instance, either with the elastic modes of the linearized system, either with the elastic modes of the underlying linear system, or with the POD (Proper Orthogonal Decomposition) modes of the nonlinear system). It should be noted that the POD modes depend on the external load vector,$\mathbb{f}(t; \mathbf{x})$. Below, it is assumed that the elastic modes of the underlying linear system are used (Sampaio and Soize 2007b). In such a framework, there are two main possibilities to construct the reduced mean computational model.

(1) The first one consists in solving the generalized eigenvalue problem associated with the mean mass and stiffness matrices for $\mathbf{x}$ fixed to its nominal value $\underline{\mathbf{x}}$. We then obtain the elastic modes of the nominal mean computational model which are independent of $\mathbf{x}$ and which depend only on $\underline{\mathbf{x}}$ which is fixed. In this case, when $\mathbf{x}$ runs through $\mathcal{C}_{\text{par}}$, matrices $[\mathbb{M}(\mathbf{x})]$ and $[\mathbb{K}(\mathbf{x})]$ have to be projected on the subspace spanned by the elastic modes of the nominal mean computational model. For very large computational model (m can be several tens of millions) such an operation is not easy to perform with usual commercial softwares which often are black boxes.

(2) The second one consists in solving the generalized eigenvalue problem associated with the mean mass and stiffness matrices for each required $\mathbf{x}$ belonging to $\mathcal{C}_{\text{par}}$. In this case, the elastic modes of the mean computational model depend on $\mathbf{x}$. In the context of the parametric probabilistic approach of model-parameter uncertainties, we then have to solve a random generalized eigenvalue problem and such an approach is better adapted to usual commercial softwares and allows a fast convergence to be obtained with respect to the dimension of the reduced-order

model. In this context, some algorithms have been developed for random eigenvalue problems of large systems (Szekely and Schuller 2001, Pradlwarter et al. 2002). In order to limit the developments, we will focus the presentation using this second approach. The extension to the first approach is straightforward from a theoretical point of view (see for instance (Soize 2000)). Finally, it should be noted that the random generalized eigenvalue problem can also be considered in a polynomial chaos decomposition for which an efficient approach has been proposed (Ghanem and Ghosh 2007). Such an ingredient can be added without difficulty in the developments presented below but would induce an addition degree of difficulty in the understanding which could mask the main ideas of the parametric probabilistic approach of model-parameter uncertainties.

For each value of $\mathbf{x}$ given in $\mathcal{C}_{\rm par}$, the generalized eigenvalue problem associated with the mean mass and stiffness matrices is written as

$$[\,\mathbb{K}(\mathbf{x})]\,\boldsymbol{\phi}(\mathbf{x}) = \lambda(\mathbf{x})\,[\,\mathbb{M}(\mathbf{x})]\,\boldsymbol{\phi}(\mathbf{x}), \tag{3.2}$$

for which the eigenvalues $0 < \lambda_1(\mathbf{x}) \leq \lambda_2(\mathbf{x}) \leq \ldots \leq \lambda_m(\mathbf{x})$ and the associated elastic modes $\{\boldsymbol{\phi}_1(\mathbf{x}), \boldsymbol{\phi}_2(\mathbf{x}), \ldots\}$ are such that

$$<[\,\mathbb{M}(\mathbf{x})]\,\boldsymbol{\phi}_\alpha(\mathbf{x})\,,\boldsymbol{\phi}_\beta(\mathbf{x})>= \mu_\alpha(\mathbf{x})\,\delta_{\alpha\beta}, \tag{3.3}$$

$$<[\,\mathbb{K}(\mathbf{x})]\,\boldsymbol{\phi}_\alpha(\mathbf{x})\,,\boldsymbol{\phi}_\beta(\mathbf{x})>= \mu_\alpha(\mathbf{x})\,\omega_\alpha(\mathbf{x})^2\,\delta_{\alpha\beta}, \tag{3.4}$$

in which $\omega_\alpha(\mathbf{x}) = \sqrt{\lambda_\alpha(\mathbf{x})}$ is the eigenfrequency of elastic mode $\boldsymbol{\phi}_\alpha(\mathbf{x})$ whose normalization is defined by the generalized mass $\mu_\alpha(\mathbf{x})$ and where $<\mathbf{u}\,,\mathbf{v}>= \sum_j u_j v_j$ is the Euclidean inner product of the vectors $\mathbf{u}$ and $\mathbf{v}$. For each value of $\mathbf{x}$ given in $\mathcal{C}_{\rm par}$, the reduced mean computational model of the dynamical system is obtained in constructing the projection of the mean computational model on the subspace $\mathbb{H}_n$ of $\mathbb{R}^m$ spanned by $\{\boldsymbol{\phi}_1(\mathbf{x}), \ldots, \boldsymbol{\phi}_n(\mathbf{x})\}$ with $n \ll m$. Let $[\,\phi(\mathbf{x})]$ be the $(m \times n)$ real matrix whose columns are vectors $\{\boldsymbol{\phi}_1(\mathbf{x}), \ldots, \boldsymbol{\phi}_n(\mathbf{x})\}$. The generalized force $\mathbf{f}(t;\mathbf{x})$ is a $\mathbb{R}^n$-vector such that $\mathbf{f}(t;\mathbf{x}) = [\,\phi(\mathbf{x})]^T\,\mathbb{f}(t;\mathbf{x})$. For all $\mathbf{x}$ in $\mathcal{C}_{\rm par}$, the generalized mass, damping and stiffness matrices $[M(\mathbf{x})]$, $[D(\mathbf{x})]$ and $[K(\mathbf{x})]$ belong to the set $\mathbb{M}_n^+(\mathbb{R})$ of all the positive-definite symmetric $(n \times n)$ real matrices, and are defined by

$$[M(\mathbf{x})]_{\alpha\beta} = \mu_\alpha(\mathbf{x})\,\delta_{\alpha\beta}, \quad [D(\mathbf{x})]_{\alpha\beta} =<[\,\mathbb{D}(\mathbf{x})]\,\boldsymbol{\phi}_\beta(\mathbf{x})\,,\boldsymbol{\phi}_\alpha(\mathbf{x})>, \tag{3.5}$$

$$[K(\mathbf{x})]_{\alpha\beta} = \mu_\alpha(\mathbf{x})\,\omega_\alpha(\mathbf{x})^2\,\delta_{\alpha\beta}, \tag{3.6}$$

in which, generally, $[D(\mathbf{x})]$ is a full matrix. Consequently, for all $\mathbf{x}$ in $\mathcal{C}_{\rm par}$ and for all fixed t, the reduced mean computational model of the dynamical system is written as the projection $\mathbf{y}^n(t)$ of $\mathbb{y}(t)$ on $\mathbb{H}_n$ which can be written as $\mathbf{y}^n(t) = [\,\phi(\mathbf{x})]\,\mathbf{q}(t)$ in which $\mathbf{q}(t)$ is the vector in $\mathbb{R}^n$ of the generalized coordinates. The reduced-order model is then written as

$$\mathbf{y}^n(t) = [\,\phi(\mathbf{x})]\,\mathbf{q}(t),$$
$$[M(\mathbf{x})]\,\ddot{\mathbf{q}}(t) + [D(\mathbf{x})]\,\dot{\mathbf{q}}(t) + [K(\mathbf{x})]\,\mathbf{q}(t) + \mathbf{F}_{\mathrm{NL}}(\mathbf{q}(t),\dot{\mathbf{q}}(t);\mathbf{x}) = \mathbf{f}(t;\mathbf{x}), \tag{3.7}$$

in which $\mathbf{F}_{\mathrm{NL}}(\mathbf{q}(t),\dot{\mathbf{q}}(t);\mathbf{x}) = [\,\phi(\mathbf{x})]^T\,\mathbb{f}_{\mathrm{NL}}([\,\phi(\mathbf{x})]\,\mathbf{q}(t),[\,\phi(\mathbf{x})]\,\dot{\mathbf{q}}(t);\mathbf{x})$. In the particular case for which $\mathbb{f}_{\mathrm{NL}} = 0$, then the corresponding equation in the frequency domain is written as

$$-\omega^2[M(\mathbf{x})]\,\mathbf{q}(\omega) + i\omega[D(\mathbf{x})]\,\mathbf{q}(\omega) + [K(\mathbf{x})]\,\mathbf{q}(\omega) = \mathbf{f}(\omega;\mathbf{x}),$$

in which $\mathbf{q}(\omega) = \int_{\mathbb{R}} e^{-i\omega t}\mathbf{q}(t)\,dt$ and $\mathbf{f}(\omega;\mathbf{x}) = \int_{\mathbb{R}} e^{-i\omega t}\mathbf{f}(t;\mathbf{x})\,dt$.

Convergence. Below, we will denote by n_0 the value of n for which the response $\mathbf{y}^n$ is converged to $\mathbb{y}$, with a given accuracy, for all the values of $\mathbf{x}$ in $\mathcal{C}_{\mathrm{par}}$.

3.3 Methodology for the parametric probabilistic approach of model-parameter uncertainties

The value of n is fixed to the value n_0 previously defined. The methodology for the parametric probabilistic approach of model-parameter uncertainties consists in modeling the uncertain model parameter $\mathbf{x}$ (whose nominal value is $\underline{\mathbf{x}}$) by a random variable $\mathbf{X}$ defined on a probability space $(\Theta,\mathcal{T},\mathcal{P})$, with values in $\mathbb{R}^{n_p}$. Consequently, the generalized matrices in Eq. (3.7) become random matrices $[M(\mathbf{X})]$, $[D(\mathbf{X})]$ and $[K(\mathbf{X})]$ and, for all fixed t, the generalized external force $\mathbf{f}(t;\mathbf{x})$ becomes a random vector $\mathbf{f}(t;\mathbf{X})$. The mean values of these random matrices are denoted by $[\underline{M}]$, $[\underline{D}]$, $[\underline{K}]$, and are such that

$$E\{[M(\mathbf{X})]\} = [\underline{M}],\;\; E\{[D(\mathbf{X})]\} = [\underline{D}],\;\; E\{[K(\mathbf{X})]\} = [\underline{K}], \tag{3.8}$$

in which E is the mathematical expectation. It should be noted that the mean matrices $[\underline{M}]$, $[\underline{D}]$ and $[\underline{K}]$ are different from the matrices $[M(\underline{\mathbf{x}})]$, $[D(\underline{\mathbf{x}})]$ and $[K(\underline{\mathbf{x}})]$ of the mean (nominal) computational model. The parametric probabilistic approach of uncertainties then consists in replacing the mean computational model by the following stochastic reduced-order computational model,

$$\mathbf{Y}(t) = [\,\phi(\mathbf{X})]\,\mathbf{Q}(t), \tag{3.9}$$

$$[M(\mathbf{X})]\,\ddot{\mathbf{Q}}(t) + [D(\mathbf{X})]\,\dot{\mathbf{Q}}(t) + [K(\mathbf{X})]\,\mathbf{Q}(t) + \mathbf{F}_{\rm NL}(\mathbf{Q}(t),\dot{\mathbf{Q}}(t);\mathbf{X}) = \mathbf{f}(t;\mathbf{X}), \tag{3.10}$$

in which for all fixed t, $\mathbf{Y}(t)$ is a $\mathbb{R}^m$-valued random vector and $\mathbf{Q}(t)$ is a $\mathbb{R}^n$-valued random vector. As soon as the probability model of random vector $\mathbf{X}$ is constructed, the stochastic computational model defined by Eqs. (3.9) and (3.10) can be solved using the methods presented in Section 2.6. In particular, the direct Monte Carlo method is efficient for a reduced-order stochastic computational model and its convergence rate is independent of the value of n_p. Consequently, such a method can be used in high dimension.

3.4 Construction of the prior probability model of model-parameter uncertainties

The unknown probability distribution of $\mathbf{X}$ is assumed to be defined by a probability density function $\mathbf{x} \mapsto p_{\mathbf{X}}(\mathbf{x})$ from $\mathbb{R}^{n_p}$ into $\mathbb{R}^+ = [0\,,+\infty[$. Under the assumption that no experimental data are available to construct $p_{\mathbf{X}}$, the prior model can be constructed using the maximum entropy principle (see Section 2.4). For such a construction, the available information has to be defined. Since $\mathbf{x}$ belongs to $\mathcal{C}_{\rm par}$, the support of $p_{\mathbf{X}}$ must be $\mathcal{C}_{\rm par}$ and the normalization condition must be verified. We then have,

$$\text{supp}\; p_{\mathbf{X}} = \mathcal{C}_{\rm par} \subset \mathbb{R}^{n_p}, \qquad \int_{\mathbb{R}^{n_p}} p_{\mathbf{X}}(\mathbf{x})\, d\mathbf{x} = \int_{\mathcal{C}_{\rm par}} p_{\mathbf{X}}(\mathbf{x})\, d\mathbf{x} = 1. \tag{3.11}$$

Since the nominal value of $\mathbf{x}$ is $\underline{\mathbf{x}} \in \mathcal{C}_{\rm par}$, an additional available information consists in writing that the mean value $E\{\mathbf{X}\}$ of $\mathbf{X}$ is equal to $\underline{\mathbf{x}}$ which yields the following constraint equation,

$$\int_{\mathbb{R}^{n_p}} \mathbf{x}\; p_{\mathbf{X}}(\mathbf{x})\, d\mathbf{x} = \underline{\mathbf{x}}. \tag{3.12}$$

In general, an additional available information can be deduced from the analysis of the mathematical properties of the solution of the stochastic computational model under construction. The random solution $\mathbf{Q}$ of the stochastic computational model defined by Eq. (3.10) must be a second-order vector-valued stochastic process (because the dynamical system is stable) which means that, for all t, we must have $E\{\|\mathbf{Q}(t)\|^2\} < +\infty$. In order that such a property be verified, it is necessary to introduce a constraint which can always be written as the equation $E\{\mathbf{h}(\mathbf{X})\} = \boldsymbol{\gamma}$ on $\mathbb{R}^{\mu_X}$, in which $\boldsymbol{\gamma} = (\gamma_1, \ldots, \gamma_{\mu_X})$ is a given vector in $\mathbb{R}^{\mu_X}$ with $\mu_X \geq 1$ and where $\mathbf{x} \mapsto \mathbf{h}(\mathbf{x}) = (h_1(\mathbf{x}), \ldots, h_{\mu_X}(\mathbf{x}))$ is a given (measurable) mapping from $\mathbb{R}^{n_p}$ into $\mathbb{R}^{\mu_X}$. Consequently, this constraint is defined as follows,

$$\int_{\mathbb{R}^{n_p}} \mathbf{h}(\mathbf{x})\, p_{\mathbf{X}}(\mathbf{x})\, d\mathbf{x} = \boldsymbol{\gamma}. \tag{3.13}$$

Let $\mathcal{C}$ be the set of all the probability density functions $p_{\mathbf{X}}$ defined on $\mathbb{R}^{n_p}$ with values in $\mathbb{R}^+$ such that Eqs. (3.11) to (3.13) hold. The prior model $p_{\mathbf{X}} \in \mathcal{C}$ can then be constructed using the maximum entropy principle (see Section 2.4). Since Eq. (3.13) is a vectorial equation of dimension μ_X, the solution $p_{\mathbf{X}}$ of the maximum entropy principle depends on the free $\mathbb{R}^{\mu_X}$-valued parameter $\boldsymbol{\gamma}$. However, parameter $\boldsymbol{\gamma}$ has generally no physical meaning and it is better to express $\boldsymbol{\gamma}$ in terms of a $\mathbb{R}^{\mu_X}$-valued parameter $\boldsymbol{\delta}_X$ which corresponds to a well defined statistical quantity for random variable $\mathbf{X}$. In general, $\boldsymbol{\delta}_X$ does not run through $\mathbb{R}^{\mu_X}$ but must belong to an admissible set $\mathcal{C}_X$ which is a subset of $\mathbb{R}^{\mu_X}$. Consequently, $p_{\mathbf{X}}$ depends on $\underline{\mathbf{x}}$ and $\boldsymbol{\delta}_X$ and we will write this probability density function as

$$\mathbf{x} \mapsto p_{\mathbf{X}}(\mathbf{x}; \underline{\mathbf{x}}, \boldsymbol{\delta}_X) \quad \text{with} \quad (\underline{\mathbf{x}}, \boldsymbol{\delta}_X) \in \mathcal{C}_{\text{par}} \times \mathcal{C}_X \subset \mathbb{R}^{n_p} \times \mathbb{R}^{\mu_X}. \tag{3.14}$$

3.5 Estimation of the parameters of the prior probability model of the uncertain model parameter

The value of n is fixed to the value n_0 for which, for all values of $\mathbf{x}$ in $\mathcal{C}_{\text{par}}$, the response $\mathbf{y}^n$ is converged with respect to n (with a given accuracy). The prior probability model $p_{\mathbf{X}}(\mathbf{x}; \underline{\mathbf{x}}, \boldsymbol{\delta}_X)$ of random variable $\mathbf{X}$ relative to the model-parameter uncertainties, depends on parameters $\underline{\mathbf{x}}$ and $\boldsymbol{\delta}_X$ belonging to admissible sets $\mathcal{C}_{\text{par}}$ and $\mathcal{C}_X$. If no experimental data are available, then $\underline{\mathbf{x}}$ is fixed to the nominal value and $\boldsymbol{\delta}_X$ must be considered as a parameter to perform a sensitivity analysis of the stochastic solution. Such a prior probability model of model-parameter uncertainties then allows the robustness of the solution to be analyzed in function of the level of model-parameter uncertainties controlled by $\boldsymbol{\delta}_X$.

For the particular case for which a few experimental data exist, $\underline{\mathbf{x}}$ can be updated and $\boldsymbol{\delta}_X$ can be estimated with the experimental data. The updating of $\underline{\mathbf{x}}$ and the estimation of $\boldsymbol{\delta}_X$ must then be performed with observations of the systems for which experimental data are available. Let $\mathbf{W}$ be the random vector which is observed and which is with values in $\mathbb{R}^{m_{\text{obs}}}$. It is assumed that $\mathbf{W}$ is independent of t but depends on $\{\mathbf{Y}(t), t \in \mathcal{J}\}$ in which $\mathcal{J}$ is any part of $\mathbb{R}$. For all $(\underline{\mathbf{x}}, \boldsymbol{\delta}_X) \in \mathcal{C}_{\text{par}} \times \mathcal{C}_X$, and for all $\mathbf{w}$ fixed, the probability density function of $\mathbf{W}$ is denoted by $p_{\mathbf{W}}(\mathbf{w};\underline{\mathbf{x}},\delta_X)$. On the other hand, it is assumed that ν_{exp} independent experimental values $\mathbf{w}_1^{\text{exp}},\ldots,\mathbf{w}_{\nu_{\text{exp}}}^{\text{exp}}$ are available. The optimal value $(\underline{\mathbf{x}}^{\text{opt}},\delta_X^{\text{opt}})$ of $(\underline{\mathbf{x}},\delta_X)$ can be estimated by maximizing the logarithm of the likelihood function (maximum likelihood method (Serfling 1980, Spall 2003)),

$$(\underline{\mathbf{x}}^{\text{opt}},\delta_X^{\text{opt}}) = \arg \max_{(\underline{\mathbf{x}},\delta_X)\in\mathcal{C}_{\text{par}}\times\mathcal{C}_X} \{\sum_{r=1}^{\nu_{\text{exp}}} \log p_{\mathbf{W}}(\mathbf{w}_r^{\text{exp}};\underline{\mathbf{x}},\delta_X)\}. \tag{3.15}$$

For all r, the quantity $p_{\mathbf{W}}(\mathbf{w}_r^{\text{exp}};\underline{\mathbf{x}},\delta_X)$ is estimated with the stochastic computational model (using, for instance, the independent realizations of $\mathbf{W}$ calculated with the Monte Carlo method and the multivariate Gaussian kernel density estimation method described in Section 8.6.5).

3.6 Posterior probability model of uncertainties using output-prediction-error method and the Bayesian method

Let $p_{\mathbf{X}}^{\text{prior}}(\mathbf{x}) = p_{\mathbf{X}}(\mathbf{x};\underline{\mathbf{x}}^{\text{opt}},\delta_X^{\text{opt}})$ be the optimal prior probability density function of $\mathbf{X}$, constructed with the optimal value $(\underline{\mathbf{x}}^{\text{opt}},\delta_X^{\text{opt}})$ of $(\underline{\mathbf{x}},\delta_X)$ which has been calculated in Section 3.5 solving the optimization problem defined by Eq. (3.15). The objective of this section is to estimate the posterior probability density function, $p_{\mathbf{X}}^{\text{post}}(\mathbf{x})$, of $\mathbf{X}$ using the experimental data associated with observation $\mathbf{W}$ introduced in Section 3.5 and using the Bayesian method.

In order to apply the Bayesian method (Serfling 1980, Bernardo and Smith 2000, Spall 2003, Congdon 2007, Carlin and Louis 2009) to estimate the posterior probability density function $p_{\mathbf{X}}^{\text{post}}(\mathbf{x})$ of $\mathbf{X}$, the output-prediction-error method (Walter and Pronzato 1997, Beck and Katafygiotis 1998, Kaipio and Somersalo 2005) is used and then, an additive noise $\mathbf{B}$ is added to the observation as explained in Section 2.2-(i). In this condition, the random observed output $\mathbf{W}^{\text{out}}$ with values in $\mathbb{R}^{m_{\text{obs}}}$, for which experimental data $\mathbf{w}_1^{\text{exp}},\ldots,\mathbf{w}_{\nu_{\text{exp}}}^{\text{exp}}$ are available (see Section 3.5), is written as

$$\mathbf{W}^{\text{out}} = \mathbf{W} + \mathbf{B}, \tag{3.16}$$

in which $\mathbf{W}$ is the computational model output with values in $\mathbb{R}^{m_{\text{obs}}}$. The noise $\mathbf{B}$ is a $\mathbb{R}^{m_{\text{obs}}}$-valued random vector, defined on probability space $(\Theta,\mathcal{T},\mathcal{P})$, which is assumed to be independent of $\mathbf{X}$, and consequently, which is independent of $\mathbf{W}$. It should be noted that this hypothesis concerning the independence of $\mathbf{W}$ and $\mathbf{B}$ could easily be removed. It is also assumed that the probability density function $p_{\mathbf{B}}(\mathbf{b})$ of random vector

$\mathbf{B}$, with respect to $d\mathbf{b}$, is known. For instance, it is often assumed that $\mathbf{B}$ is a centered Gaussian random vector for which the covariance matrix $[C_{\mathbf{B}}] = E\{\mathbf{B}\mathbf{B}^T\}$ is assumed to be invertible. In such a case, we would have

$$p_{\mathbf{B}}(\mathbf{b}) = (2\pi)^{-m_{\rm obs}/2} \, (\det[C_{\mathbf{B}}])^{-1/2} \, \exp(-\frac{1}{2} < [C_{\mathbf{B}}]^{-1}\mathbf{b} \, , \mathbf{b} >). \quad (3.17)$$

The posterior probability density function $p_{\mathbf{X}}^{\rm post}(\mathbf{x})$ is then calculated by the Bayes formula,

$$p_{\mathbf{X}}^{\rm post}(\mathbf{x}) = \mathcal{L}(\mathbf{x}) \, p_{\mathbf{X}}^{\rm prior}(\mathbf{x}), \quad (3.18)$$

in which $\mathbf{x} \mapsto \mathcal{L}(\mathbf{x})$ is the likelihood function defined on $\mathbb{R}^{n_p}$, with values in $\mathbb{R}^+$, such that

$$\mathcal{L}(\mathbf{x}) = \frac{\Pi_{r=1}^{\nu_{\rm exp}} \, p_{\mathbf{W}^{\rm out}|\mathbf{X}}(\mathbf{w}_r^{\rm exp}|\mathbf{x})}{E\{\Pi_{r=1}^{\nu_{\rm exp}} \, p_{\mathbf{W}^{\rm out}|\mathbf{X}}(\mathbf{w}_r^{\rm exp}|\mathbf{X}^{\rm prior})\}} \, . \quad (3.19)$$

In Eq. (3.19), $p_{\mathbf{W}^{\rm out}|\mathbf{X}}(\mathbf{w}_\ell^{\rm exp}|\mathbf{x})$ is the experimental value of the conditional probability density function $\mathbf{w} \mapsto p_{\mathbf{W}^{\rm out}|\mathbf{X}}(\mathbf{w}|\mathbf{x})$ of the random observed output $\mathbf{W}^{\rm out}$ given $\mathbf{X} = \mathbf{x}$ in $\mathcal{C}_{\rm par}$. Taking into account Eq. (3.16) and that $\mathbf{B}$ and $\mathbf{W}$ are assumed to be independent, it can easily be deduced that

$$p_{\mathbf{W}^{\rm out}|\mathbf{X}}(\mathbf{w}_r^{\rm exp}|\mathbf{x}) = p_{\mathbf{B}}(\mathbf{w}_r^{\rm exp} - \mathbb{h}(\mathbf{x})), \quad (3.20)$$

in which $\mathbf{w} = \mathbb{h}(\mathbf{x})$ is the model observation depending on $\{\mathbf{y}^n(t), t \in \mathcal{J}\}$ which is calculated using Eq. (3.7). It should be noted that the posterior probability density function strongly depends on the choice of the probability model of the output additive noise $\mathbf{B}$.

Chapter 4

Nonparametric probabilistic approach to uncertainties in computational structural dynamics

The nonparametric probabilistic approach of uncertainties has been introduced in (Soize 2000; 2001; 2005b) to take into account both the model-parameter uncertainties and the model uncertainties induced by modeling errors, without separating the contribution of each one of these two types of uncertainties. Below, we summarized this approach which allows the uncertainties to be taken into account in the computational model.

4.1 Methodology to take into account both the model-parameter uncertainties and the model uncertainties (modeling errors)

Let $(\Theta', \mathcal{T}', \mathcal{P}')$ be another probability space (we recall that probability space $(\Theta, \mathcal{T}, \mathcal{P})$ is devoted to the probabilistic model of model-parameter uncertainties using the parametric probabilistic approach (Section 3). Let $\underline{\mathbf{x}}$ be the nominal value of model parameter $\mathbf{x}$, which is fixed in $\mathcal{C}_{\text{par}}$. Let $\underline{\mathbf{y}}^n$ be the response of the mean reduced-order computational model at order n for the nominal value $\underline{\mathbf{x}}$ of parameter $\mathbf{x}$, and which is such that (see Eq. (3.7)),

$$\underline{\mathbf{y}}^n(t) = [\,\phi(\underline{\mathbf{x}})]\,\underline{\mathbf{q}}(t), \tag{4.1}$$

$$[M(\underline{\mathbf{x}})]\ddot{\underline{\mathbf{q}}}(t)+[D(\underline{\mathbf{x}})]]\dot{\underline{\mathbf{q}}}(t)+[K(\underline{\mathbf{x}})]]\underline{\mathbf{q}}(t)+\mathbf{F}_{\text{NL}}(\underline{\mathbf{q}}(t),\dot{\underline{\mathbf{q}}}(t);\underline{\mathbf{x}})=\mathbf{f}(t;\underline{\mathbf{x}}). \tag{4.2}$$

The value of n is fixed to the value n_0 defined at the end of Section 3.2, for which, for all $\underline{\mathbf{x}}$ fixed in $\mathcal{C}_{\mathrm{par}}$, the response $\underline{\mathbf{y}}^n$ is converged with respect to n. For n fixed to the value n_0, the nonparametric probabilistic approach of uncertainties then consists in replacing, in Eq. (4.2), the matrices $[M(\underline{\mathbf{x}})]$, $[D(\underline{\mathbf{x}})]$ and $[K(\underline{\mathbf{x}})]$ by the random matrices

$$
\begin{aligned}
[\mathbf{M}(\underline{\mathbf{x}})] &= \{\theta' \mapsto [\mathbf{M}(\theta';\underline{\mathbf{x}})]\}, \\
[\mathbf{D}(\underline{\mathbf{x}})] &= \{\theta' \mapsto [\mathbf{D}(\theta';\underline{\mathbf{x}})]\}, \qquad (4.3)\\
[\mathbf{K}(\underline{\mathbf{x}})] &= \{\theta' \mapsto [\mathbf{K}(\theta';\underline{\mathbf{x}})]\}, \qquad (4.4)
\end{aligned}
$$

defined on probability space $(\Theta', \mathcal{T}', \mathcal{P}')$, depending on the nominal value $\underline{\mathbf{x}}$ of parameter $\mathbf{x}$ and for which the probability distributions will be defined in Section 4.2. Taking into account such a construction, it can be proven that

$$
E\{[\mathbf{M}(\underline{\mathbf{x}})]\} = [M(\underline{\mathbf{x}})] \; , \; E\{[\mathbf{D}(\underline{\mathbf{x}})]\} = [D(\underline{\mathbf{x}})] \; , \; E\{[\mathbf{K}(\underline{\mathbf{x}})]\} = [K(\underline{\mathbf{x}})] \; . \qquad (4.5)
$$

The deterministic reduced computational model defined by Eqs. (4.1) and (4.2), is then replaced by the following stochastic reduced computational model,

$$
\mathbf{Y}(t) = [\phi(\underline{\mathbf{x}})]\, \mathbf{Q}(t), \qquad (4.6)
$$

$$
[\mathbf{M}(\underline{\mathbf{x}})]\, \ddot{\mathbf{Q}}(t) + [\mathbf{D}(\underline{\mathbf{x}})]\, \dot{\mathbf{Q}}(t) + [\mathbf{K}(\underline{\mathbf{x}})]\, \mathbf{Q}(t) + \mathbf{F}_{\mathrm{NL}}(\mathbf{Q}(t), \dot{\mathbf{Q}}(t); \underline{\mathbf{x}}) = \mathbf{f}(t; \underline{\mathbf{x}}), \qquad (4.7)
$$

in which for all fixed t, $\mathbf{Y}(t) = \{\theta' \mapsto \mathbf{Y}(\theta'; t)\}$ and $\mathbf{Q}(t) = \{\theta' \mapsto \mathbf{Q}(\theta'; t)\}$ are $\mathbb{R}^m$- and $\mathbb{R}^n$-valued random vectors defined on Θ'. The realizations $\mathbf{Y}(\theta'; t)$ and $\mathbf{Q}(\theta'; t)$ of the random variables $\mathbf{Y}(t)$ and $\mathbf{Q}(t)$ verify the deterministic equations

$$
\mathbf{Y}(\theta'; t) = [\phi(\underline{\mathbf{x}})]\, \mathbf{Q}(\theta'; t), \qquad (4.8)
$$

$$
\begin{aligned}
&[\mathbf{M}(\theta';\underline{\mathbf{x}})]\, \ddot{\mathbf{Q}}(\theta'; t) + [\mathbf{D}(\theta';\underline{\mathbf{x}})]\, \dot{\mathbf{Q}}(\theta'; t) + [\mathbf{K}(\theta';\underline{\mathbf{x}})]\, \mathbf{Q}(\theta'; t) \\
&\qquad + \mathbf{F}_{\mathrm{NL}}(\mathbf{Q}(\theta'; t), \dot{\mathbf{Q}}(\theta'; t); \underline{\mathbf{x}}) = \mathbf{f}(t; \underline{\mathbf{x}})\, . \qquad (4.9)
\end{aligned}
$$

4.2 Construction of the prior probability model of the random matrices

For n fixed to the value n_0 and as explained in (Soize 2000; 2001; 2005b), the random matrices $[\mathbf{M}(\underline{\mathbf{x}})]$, $[\mathbf{D}(\underline{\mathbf{x}})]$ and $[\mathbf{K}(\underline{\mathbf{x}})]$ introduced in Eq. (4.7) are written as

$$
\begin{aligned}
[\mathbf{M}(\underline{\mathbf{x}})] &= [L_M(\underline{\mathbf{x}})]^T\, [\mathbf{G}_M]\, [L_M(\underline{\mathbf{x}})], \\
[\mathbf{D}(\underline{\mathbf{x}})] &= [L_D(\underline{\mathbf{x}})]^T\, [\mathbf{G}_D]\, [L_D(\underline{\mathbf{x}})], \qquad (4.10)\\
[\mathbf{K}(\underline{\mathbf{x}})] &= [L_K(\underline{\mathbf{x}})]^T\, [\mathbf{G}_K]\, [L_K(\underline{\mathbf{x}})],
\end{aligned}
$$

in which $[L_M(\underline{\mathbf{x}})]$, $[L_D(\underline{\mathbf{x}})]$ and $[L_K(\underline{\mathbf{x}})]$ are the upper triangular matrices such that $[M(\underline{\mathbf{x}})] = [L_M(\underline{\mathbf{x}})]^T\,[L_M(\underline{\mathbf{x}})]$, $[D(\underline{\mathbf{x}})] = [L_D(\underline{\mathbf{x}})]^T\,[L_D(\underline{\mathbf{x}})]$ and $[K(\underline{\mathbf{x}})] = [L_K(\underline{\mathbf{x}})]^T\,[L_K(\underline{\mathbf{x}})]$.

In Eq. (4.10), $[\mathbf{G}_M]$, $[\mathbf{G}_D]$ and $[\mathbf{G}_K]$ are random matrices which are defined on probability space $(\Theta', \mathcal{T}', \mathcal{P}')$ with values in $\mathbb{M}_n^+(\mathbb{R})$. The joint probability density function of these random matrices $[\mathbf{G}_M]$, $[\mathbf{G}_D]$ and $[\mathbf{G}_K]$ is constructed using the maximum entropy principle under the constraints defined by the available information. Taking into account the available information introduced in the nonparametric probabilistic approach (Soize 2000; 2001; 2005b), it is proven that these three random matrices are statistically independent and each one belongs to ensemble $\mathrm{SG}_\varepsilon^+$ of random matrices defined in Section 2.5.3. Consequently, these three random matrices depend on the free positive real dispersion parameters δ_M, δ_D and δ_K which allow the level of the statistical fluctuations, that is to say the level of uncertainties, to be controlled. Let $\boldsymbol{\delta}_G = (\delta_M, \delta_D, \delta_K)$ be the vector of the dispersion parameters, which belongs to an admissible set $\mathcal{C}_G \subset \mathbb{R}^3$. Consequently, the joint probability density function of the random matrices $[\mathbf{G}_M]$, $[\mathbf{G}_D]$ and $[\mathbf{G}_K]$ is written as

$$
\begin{aligned}
([G_M], [G_D], [G_K]) \mapsto p_{\mathbf{G}}([G_M], [G_D], [G_K]; \boldsymbol{\delta}_G) &= p_{[\mathbf{G}_M]]}([G_M]; \delta_M) \\
\times\, p_{[\mathbf{G}_D]]}([G_D]; \delta_D) \,\times\, p_{[\mathbf{G}_K]]}([G_K]; \delta_K), &\quad \boldsymbol{\delta}_G \in \mathcal{C}_G \subset \mathbb{R}^3. \qquad (4.11)
\end{aligned}
$$

The probability distributions of random matrices $[\mathbf{G}_M]$, $[\mathbf{G}_D]$ and $[\mathbf{G}_K]$ and their algebraic representations, which are useful for generating independent realizations $[\mathbf{G}_M(\theta')]$, $[\mathbf{G}_D(\theta')]$ and $[\mathbf{G}_K(\theta')]$, are explicitly defined in Section 2.5.3. From Eq. (4.10), for θ' in Θ', it can then be deduced the realizations $[\mathbf{M}(\theta', \underline{\mathbf{x}})]$, $[\mathbf{D}(\theta', \underline{\mathbf{x}})]$ and $[\mathbf{K}(\theta', \underline{\mathbf{x}})]$ which are given by

$$
\begin{aligned}
[\mathbf{M}(\theta'; \underline{\mathbf{x}})] &= [L_M(\underline{\mathbf{x}})]^T\,[\mathbf{G}_M(\theta')]\,[L_M(\underline{\mathbf{x}})], \\
[\mathbf{D}(\theta'; \underline{\mathbf{x}})] &= [L_D(\underline{\mathbf{x}})]^T\,[\mathbf{G}_D(\theta')]\,[L_D(\underline{\mathbf{x}})], \\
[\mathbf{K}(\theta'; \underline{\mathbf{x}})] &= [L_K(\underline{\mathbf{x}})]^T\,[\mathbf{G}_K(\theta')]\,[L_K(\underline{\mathbf{x}})].
\end{aligned}
$$

4.3 Estimation of the parameters of the prior probability model of uncertainties

As explained above, in the nonparametric probabilistic approach of uncertainties, n is fixed to the value n_0 for which the response of the mean reduced-order computational model is converged with respect to n. The prior probability model of uncertainties then depends on parameters $\underline{\mathbf{x}}$ and $\boldsymbol{\delta}_G$ belonging to the admissible sets $\mathcal{C}_{\text{par}}$ and $\mathcal{C}_G$. If no experimental data are available, then $\underline{\mathbf{x}}$ is fixed to its nominal value and $\boldsymbol{\delta}_G$ must be considered as a vector-valued parameter for performing a sensitivity analysis of the stochastic solution with respect to the level of uncertainties. Such a prior probability model of both the model-parameter uncertainties and the model uncertainties then allows the robustness of the solution to be analyzed as a function of the level of uncertainties which is controlled by $\boldsymbol{\delta}_G$.

If a few experimental data are available, the methodology presented in Section 3.5 can then be used to update the nominal value $\underline{\mathbf{x}}$ and to estimate $\boldsymbol{\delta}_G$. As previously, let $\mathbf{W}$ be the random vector which is observed, which is independent of t, but which depends on $\{\mathbf{Y}(t), t \in \mathcal{J}\}$, in which $\mathcal{J}$ is any part of $\mathbb{R}$ and where $\mathbf{Y}$ is the vector-valued stochastic process which is the second-order random solution of Eqs. (4.6) and (4.7). For all $(\underline{\mathbf{x}}, \boldsymbol{\delta}_G)$ in $\mathcal{C}_{\text{par}} \times \mathcal{C}_G$, the probability density function of $\mathbf{W}$ is denoted by $\mathbf{w} \mapsto p_{\mathbf{W}}(\mathbf{w}; \underline{\mathbf{x}}, \boldsymbol{\delta}_G)$. The optimal value $(\underline{\mathbf{x}}^{\text{opt}}, \boldsymbol{\delta}_G^{\text{opt}})$ of $(\underline{\mathbf{x}}, \boldsymbol{\delta}_G)$ can be estimated by maximizing the logarithm of the likelihood function,

$$(\underline{\mathbf{x}}^{\text{opt}}, \boldsymbol{\delta}_G^{\text{opt}}) = \arg \max_{(\underline{\mathbf{x}}, \boldsymbol{\delta}_G) \in \mathcal{C}_{\text{par}} \times \mathcal{C}_G} \{ \sum_{r=1}^{\nu_{\text{exp}}} \log p_{\mathbf{W}}(\mathbf{w}_r^{\text{exp}}; \underline{\mathbf{x}}, \boldsymbol{\delta}_G \}, \tag{4.12}$$

in which $\mathbf{w}_1^{\text{exp}}, \ldots, \mathbf{w}_{\nu_{\text{exp}}}^{\text{exp}}$ are ν_{exp} independent experimental data corresponding to $\mathbf{W}$. Similarly to Section 3.5, for all r, $p_{\mathbf{W}}(\mathbf{w}_r^{\text{exp}}; \underline{\mathbf{x}}, \boldsymbol{\delta}_G)$ is estimated with the stochastic computational model defined by Eqs. (4.6) and (4.7) (using, for instance, the independent realizations of $\mathbf{W}$ calculated with the Monte Carlo method and the multivariate Gaussian kernel density estimation method described in Section 8.6.5).

4.4 Comments about the applications and the validation of the nonparametric probabilistic approach of uncertainties

Concerning the applications in linear and nonlinear structural dynamics, we refer the reader to (Soize 2000; 2001; 2003a;b, Soize and Chebli 2003, Capiez-Lernout and Soize 2004, Desceliers et al. 2004, Capiez-Lernout et al. 2005, Arnst et al. 2006, Capiez-Lernout et al. 2006, Cottereau et al. 2007, Sampaio and Soize 2007a, Cottereau et al. 2008, Mignolet and Soize 2008a, Pellissetti et al. 2008, Ritto et al. 2009).

Identification methodologies, applied to nonlinear structural dynamical problems, can be found in (Soize et al. 2008a, Batou and Soize 2009b;a).

Experimental validations are presented in (Chebli and Soize 2004, Chen et al. 2006, Duchereau and Soize 2006, Durand et al. 2008, Desceliers et al. 2009).

Some applications devoted to robust optimization problems, that is to say to optimization problems using uncertain computational models, can be found in (Capiez-Lernout and Soize 2008b;c, Soize et al. 2008b, Ritto et al. 2010).

Chapter 5

Generalized probabilistic approach to uncertainties in computational structural dynamics

In Section 4, we have presented the nonparametric probabilistic approach of uncertainties which has been introduced in (Soize 2000; 2001; 2005b) to take into account both the model-parameter uncertainties and the model uncertainties induced by modeling errors, without separating the contribution of each one of these two types of uncertainties. An extension of the nonparametric approach of uncertainties, called the generalized probabilistic approach of uncertainties, has been proposed (Soize 2010a) and allows the prior probability model of each type of uncertainties (model-parameter uncertainties and model uncertainties induced by modeling errors) to be separately constructed. Below, we summarized this approach which is based on the use of the parametric probabilistic approach of model-parameter uncertainties, presented in Section 3, and on the use of the nonparametric probabilistic approach of model uncertainties induced by modeling errors, introduced in Section 4.

5.1 Methodology of the generalized probabilistic approach

Probability space $(\Theta, \mathcal{T}, \mathcal{P})$, introduced in Section 3, is devoted to the probability model of model-parameter uncertainties using the parametric probabilistic approach, while $(\Theta', \mathcal{T}', \mathcal{P}')$, introduced in Section 4, is devoted to the probability model of model uncertainties (modeling errors) using the nonparametric probabilistic approach. Similarly to the nonparametric probabilistic approach of uncertainties introduced in Section 4, the value of n is fixed to the value n_0 defined at the end of Section 3.2. For all $\mathbf{x}$ in $\mathcal{C}_{\text{par}}$, the response $\mathbf{y}^n$ computed with Eq. (3.7), is converged with respect to n (with a given accuracy). For n fixed to the value n_0, the generalized probabilistic approach of uncertainties then consists in replacing, for all $\mathbf{x}$ in $\mathcal{C}_{\text{par}}$, the matrices $[M(\mathbf{x})]$, $[D(\mathbf{x})]$ and $[K(\mathbf{x})]$ by the random matrices

$$
\begin{aligned}
[\mathbf{M}(\mathbf{x})] &= \{\theta' \mapsto [\mathbf{M}(\theta';\mathbf{x})]\},\\
[\mathbf{D}(\mathbf{x})] &= \{\theta' \mapsto [\mathbf{D}(\theta';\mathbf{x})]\},\\
[\mathbf{K}(\mathbf{x})] &= \{\theta' \mapsto [\mathbf{K}(\theta';\mathbf{x})]\},
\end{aligned}
$$

on probability space $(\Theta', \mathcal{T}', \mathcal{P}')$ and which will be defined in Section 5.2. The generalized probabilistic approach of uncertainties then consists in replacing in Eq. (3.10), in order to take into account modeling errors, the dependent random matrices $[M(\mathbf{X})]$, $[D(\mathbf{X})]$ and $[K(\mathbf{X})]$ by the dependent random matrices

$$
\begin{aligned}
[\mathbf{M}(\mathbf{X})] &= \{(\theta,\theta') \mapsto [\mathbf{M}(\theta';\mathbf{X}(\theta))]\},\\
[\mathbf{D}(\mathbf{X})] &= \{(\theta,\theta') \mapsto [\mathbf{D}(\theta';\mathbf{X}(\theta))]\},\\
[\mathbf{K}(\mathbf{X})] &= \{(\theta,\theta') \mapsto [\mathbf{K}(\theta';\mathbf{X}(\theta))]\},
\end{aligned}
$$

defined on $\Theta \times \Theta'$. Taking into account the construction which will be given in Section 5.2, it can easily be proven that

$$
E\{[\mathbf{M}(\mathbf{X})]\} = [\underline{M}],\ \ E\{[\mathbf{D}(\mathbf{X})]\} = [\underline{D}],\ \ E\{[\mathbf{K}(\mathbf{X})]\} = [\underline{K}], \tag{5.1}
$$

in which the matrices $[\underline{M}]$, $[\underline{D}]$ and $[\underline{K}]$ are the deterministic matrices introduced in Section 3.3. The stochastic reduced computational model, defined by Eqs. (3.9) and (3.10), is then replaced by the following stochastic reduced computational model,

$$
\mathbf{Y}(t) = [\,\phi(\mathbf{X})]\,\mathbf{Q}(t), \tag{5.2}
$$

$$
[\mathbf{M}(\mathbf{X})]\,\ddot{\mathbf{Q}}(t) + [\mathbf{D}(\mathbf{X})]\,\dot{\mathbf{Q}}(t) + [\mathbf{K}(\mathbf{X})]\,\mathbf{Q}(t) + \mathbf{F}_{\text{NL}}(\mathbf{Q}(t),\dot{\mathbf{Q}}(t);\mathbf{X}) = \mathbf{f}(t;\mathbf{X}), \tag{5.3}
$$

in which for all fixed t, $\mathbf{Y}(t) = \{(\theta,\theta') \mapsto \mathbf{Y}(\theta,\theta';t)\}$ and $\mathbf{Q}(t) = \{(\theta,\theta') \mapsto \mathbf{Q}(\theta,\theta';t)\}$ are $\mathbb{R}^m$- and $\mathbb{R}^n$-valued random vectors defined on $\Theta \times \Theta'$. Let $\mathbf{X}(\theta_\ell)$ be any realization of random variable $\mathbf{X}$ for θ_ℓ in Θ. For all $\mathbf{x}$ in $\mathcal{C}_{\text{par}}$, let $[\mathbf{M}(\theta'_{\ell'};\mathbf{x})]$, $[\mathbf{D}(\theta'_{\ell'};\mathbf{x})]$ and $[\mathbf{K}(\theta'_{\ell'};\mathbf{x})]$ be any realizations of random matrices $[\mathbf{M}(\mathbf{x})]$, $[\mathbf{D}(\mathbf{x})]$, $[\mathbf{K}(\mathbf{x})]$ for $\theta'_{\ell'}$ in Θ'. The realizations $\mathbf{Y}(\theta_\ell,\theta'_{\ell'};t)$ and $\mathbf{Q}(\theta_\ell,\theta'_{\ell'};t)$ of random variables $\mathbf{Y}(t)$ and $\mathbf{Q}(t)$ verify the following deterministic equations

$$
\mathbf{Y}(\theta_\ell,\theta'_{\ell'};t) = [\,\phi(\mathbf{X}(\theta_\ell))]\,\mathbf{Q}(\theta_\ell,\theta'_{\ell'};t), \tag{5.4}
$$

$$[\mathbf{M}(\theta'_{\ell'};\mathbf{X}(\theta_\ell))]\,\ddot{\mathbf{Q}}(\theta_\ell,\theta'_{\ell'};t)+[\mathbf{D}(\theta'_{\ell'};\mathbf{X}(\theta_\ell))]\,\dot{\mathbf{Q}}(\theta_\ell,\theta'_{\ell'};t)$$
$$+[\mathbf{K}(\theta'_{\ell'};\mathbf{X}(\theta_\ell))]\,\mathbf{Q}(\theta_\ell,\theta'_{\ell'};t)+\mathbf{F}_{\rm NL}(\mathbf{Q}(\theta_\ell,\theta'_{\ell'};t),\dot{\mathbf{Q}}(\theta_\ell,\theta'_{\ell'};t);\mathbf{X}(\theta_\ell))$$
$$=\mathbf{f}(t;\mathbf{X}(\theta_\ell)). \tag{5.5}$$

5.2 Construction of the prior probability model of the random matrices

As explained in (Soize 2010a), the dependent random matrices $[\mathbf{M}(\mathbf{X})]$, $[\mathbf{D}(\mathbf{X})]$ and $[\mathbf{K}(\mathbf{X})]$, introduced in Eq. (5.3), are written as

$$\begin{aligned}
[\mathbf{M}(\mathbf{X})]&=[L_M(\mathbf{X})]^T\,[\mathbf{G}_M]\,[L_M(\mathbf{X})],\\
[\mathbf{D}(\mathbf{X})]&=[L_D(\mathbf{X})]^T\,[\mathbf{G}_D]\,[L_D(\mathbf{X})],\\
[\mathbf{K}(\mathbf{X})]&=[L_K(\mathbf{X})]^T\,[\mathbf{G}_K]\,[L_K(\mathbf{X})],
\end{aligned} \tag{5.6}$$

in which, for all $\mathbf{x}$ in $\mathcal{C}_{\rm par}$, $[L_M(\mathbf{x})]$, $[L_D(\mathbf{x})]$ and $[L_K(\mathbf{x})]$ are the upper triangular matrices such that $[M(\mathbf{x})]=[L_M(\mathbf{x})]^T\,[L_M(\mathbf{x})]$, $[D(\mathbf{x})]=[L_D(\mathbf{x})]^T\,[L_D(\mathbf{x})]$ and $[K(\mathbf{x})]=[L_K(\mathbf{x})]^T\,[L_K(\mathbf{x})]$. In Eq. (5.6), $[\mathbf{G}_M]$, $[\mathbf{G}_D]$ and $[\mathbf{G}_K]$ are the random matrices on probability space $(\Theta',\mathcal{T}',\mathcal{P}')$, with values in $\mathbb{M}_n^+(\mathbb{R})$, defined in Section 4.2. These three random matrices are statistically independent and each one belongs to ensemble $\mathrm{SG}_\varepsilon^+$ of random matrices defined in Section 2.5.3. Consequently, these three random matrices depend on free positive real dispersion parameters δ_M, δ_D and δ_K which allow the level of model uncertainties to be controlled. As previously, let $\boldsymbol{\delta}_G=(\delta_M,\delta_D,\delta_K)$ be the vector of the dispersion parameters, which belongs to an admissible set $\mathcal{C}_G\subset\mathbb{R}^3$. The algebraic representation of each random matrix $[\mathbf{G}_M]$, $[\mathbf{G}_D]$ or $[\mathbf{G}_K]$, given in Section 2.5.3, is useful for generating independent realizations $[\mathbf{G}_M(\theta')]$, $[\mathbf{G}_D(\theta')]$ and $[\mathbf{G}_K(\theta')]$. From Eq. (5.6), for (θ,θ') in $\Theta\times\Theta'$, it can then be deduced the realization $[\mathbf{M}(\theta';\mathbf{X}(\theta))]$, $[\mathbf{D}(\theta';\mathbf{X}(\theta))]$ and $[\mathbf{K}(\theta';\mathbf{X}(\theta))]$ which are given by

$$\begin{aligned}
[\mathbf{M}(\theta';\mathbf{X}(\theta))]&=[L_M(\mathbf{X}(\theta))]^T\,[\mathbf{G}_M(\theta')]\,[L_M(\mathbf{X}(\theta))],\\
[\mathbf{D}(\theta';\mathbf{X}(\theta))]&=[L_D(\mathbf{X}(\theta))]^T\,[\mathbf{G}_D(\theta')]\,[L_D(\mathbf{X}(\theta))],\\
[\mathbf{K}(\theta';\mathbf{X}(\theta))]&=[L_K(\mathbf{X}(\theta))]^T\,[\mathbf{G}_K(\theta')]\,[L_K(\mathbf{X}(\theta))].
\end{aligned}$$

5.3 Estimation of the parameters of the prior probability model of uncertainties

The value of n is fixed to the value n_0 defined in Section 5.1. The prior probability model of uncertainties then depends on parameters $\underline{\mathbf{x}}$, $\boldsymbol{\delta}_X$ and $\boldsymbol{\delta}_G$ belonging to the admissible sets $\mathcal{C}_{\rm par}$, $\mathcal{C}_X$ and $\mathcal{C}_G$. If no experimental data are available, then $\underline{\mathbf{x}}$ is fixed to the nominal value $\mathbf{x}_0$ and, $\boldsymbol{\delta}_X$ and $\boldsymbol{\delta}_G$ must be considered as parameters to perform a sensitivity analysis of the stochastic solution. Such a prior probability model of the model-parameter uncertainties and the model uncertainties then allows the robustness of the solution to be analyzed in function of the level of model-parameter uncertainties controlled by $\boldsymbol{\delta}_X$ and in function of model uncertainties controlled by $\boldsymbol{\delta}_G$.

If a few experimental data are available, the methodology presented in Sections 3.5 and 4.3, can then be used to update $\underline{\mathbf{x}}$ and to estimate $\boldsymbol{\delta}_X$ and $\boldsymbol{\delta}_G$. As previously, let $\mathbf{W}$ be the random vector which is observed, which is independent of t, but which depends on $\{\mathbf{Y}(t), t \in \mathcal{J}\}$, in which $\mathcal{J}$ is any part of $\mathbb{R}$ and where $\mathbf{Y}$ is the vector-valued stochastic process which is the second-order random solution of Eqs. (5.2) and (5.3). For all $(\underline{\mathbf{x}}, \boldsymbol{\delta}_X, \boldsymbol{\delta}_G) \in \mathcal{C}_{\text{par}} \times \mathcal{C}_X \times \mathcal{C}_G$, the probability density function of $\mathbf{W}$ is denoted by $\mathbf{w} \mapsto p_{\mathbf{W}}(\mathbf{w}; \underline{\mathbf{x}}, \boldsymbol{\delta}_X, \boldsymbol{\delta}_G)$. The optimal value $(\underline{\mathbf{x}}^{\text{opt}}, \boldsymbol{\delta}_X^{\text{opt}}, \boldsymbol{\delta}_G^{\text{opt}})$ of $(\underline{\mathbf{x}}, \boldsymbol{\delta}_X, \boldsymbol{\delta}_G)$ can be estimated by maximizing the logarithm of the likelihood function,

$$(\underline{\mathbf{x}}^{\text{opt}}, \boldsymbol{\delta}_X^{\text{opt}}, \boldsymbol{\delta}_G^{\text{opt}}) = \arg \max_{(\underline{\mathbf{x}}, \boldsymbol{\delta}_X, \boldsymbol{\delta}_G) \in \mathcal{C}_{\text{par}} \times \mathcal{C}_X \times \mathcal{C}_G} \{ \sum_{r=1}^{\nu_{\text{exp}}} \log p_{\mathbf{W}}(\mathbf{w}_r^{\text{exp}}; \underline{\mathbf{x}}, \boldsymbol{\delta}_X, \boldsymbol{\delta}_G \}, \tag{5.7}$$

in which $\mathbf{w}_1^{\text{exp}}, \ldots, \mathbf{w}_{\nu_{\text{exp}}}^{\text{exp}}$ are ν_{exp} independent experimental data corresponding to $\mathbf{W}$. Similarly to Sections 3.5 and 4.3, for all r, the quantity $p_{\mathbf{W}}(\mathbf{w}_r^{\text{exp}}; \underline{\mathbf{x}}, \boldsymbol{\delta}_X, \boldsymbol{\delta}_G)$ is estimated with the stochastic computational model defined by Eqs. (5.2) and (5.3) (using, for instance, the independent realizations of $\mathbf{W}$ calculated with the Monte Carlo method and the multivariate Gaussian kernel density estimation method described in Section 8.6.5).

5.4 Posterior probability model of uncertainties using the Bayesian method

Let $p_{\mathbf{X}}^{\text{prior}}(\mathbf{x}) = p_{\mathbf{X}}(\mathbf{x}; \underline{\mathbf{x}}^{\text{opt}}, \boldsymbol{\delta}_X^{\text{opt}})$ and $p_{\mathbf{G}}([G_M], [G_D], [G_K]; \boldsymbol{\delta}_G^{\text{opt}})$ be the optimal prior probability density functions of random vector $\mathbf{X}$ and of random matrices $[\mathbf{G}_M]$, $[\mathbf{G}_D]$, $[\mathbf{G}_K]$, constructed with the optimal value $(\underline{\mathbf{x}}^{\text{opt}}, \boldsymbol{\delta}_X^{\text{opt}}, \boldsymbol{\delta}_G^{\text{opt}})$ of $(\underline{\mathbf{x}}, \boldsymbol{\delta}_X, \boldsymbol{\delta}_G)$ which has been calculated in Section 5.3. The objective of this section is the following. For this given optimal prior probability model $p_{\mathbf{G}}([G_M], [G_D], [G_K]; \boldsymbol{\delta}_G^{\text{opt}})$ of model uncertainties induced by modeling errors, the posterior probability density function $p_{\mathbf{X}}^{\text{post}}(\mathbf{x})$ of model-parameter uncertainties is estimated using the experimental data associated with observation $\mathbf{W}$ and using the Bayesian method.

5.4.1 Posterior probability density function of the model-parameter uncertainties

Let $\mathbf{w}_1^{\text{exp}}, \ldots, \mathbf{w}_{\nu_{\text{exp}}}^{\text{exp}}$ be the ν_{exp} independent experimental data corresponding to observation $\mathbf{W}$ introduced in Section 5.3. The Bayesian method allows the posterior probability density function $p_{\mathbf{X}}^{\text{post}}(\mathbf{x})$ to be calculated by

$$p_{\mathbf{X}}^{\text{post}}(\mathbf{x}) = \mathcal{L}(\mathbf{x})\, p_{\mathbf{X}}^{\text{prior}}(\mathbf{x}), \tag{5.8}$$

in which $\mathbf{x} \mapsto \mathcal{L}(\mathbf{x})$ is the likelihood function defined on $\mathbb{R}^{n_p}$, with values in $\mathbb{R}^+$, such that

$$\mathcal{L}(\mathbf{x}) = \frac{\Pi_{r=1}^{\nu_{\text{exp}}}\, p_{\mathbf{W}|\mathbf{X}}(\mathbf{w}_r^{\text{exp}}|\mathbf{x})}{E\{\Pi_{r=1}^{\nu_{\text{exp}}}\, p_{\mathbf{W}|\mathbf{X}}(\mathbf{w}_r^{\text{exp}}|\mathbf{X}^{\text{prior}})\}}. \tag{5.9}$$

In Eq. (5.9), $p_{\mathbf{W}|\mathbf{X}}(\mathbf{w}_\ell^{\text{exp}}|\mathbf{x})$ is the experimental value of the conditional probability density function $\mathbf{w} \mapsto p_{\mathbf{W}|\mathbf{X}}(\mathbf{w}|\mathbf{x})$ of the random observation $\mathbf{W}$ given $\mathbf{X} = \mathbf{x}$ in $\mathcal{C}_{\text{par}}$. Equation (5.9) shows that likelihood function $\mathcal{L}$ must verify the following equation,

$$E\{\mathcal{L}(\mathbf{X}^{\text{prior}})\} = \int_{\mathbb{R}^{n_p}} \mathcal{L}(\mathbf{x})\, p_{\mathbf{X}}^{\text{prior}}(\mathbf{x})\, d\mathbf{x} = 1. \tag{5.10}$$

5.4.2 Posterior probability density functions of the responses

Let $\mathbf{U} = (\mathbf{V}, \mathbf{W})$ be the random response vector in which $\mathbf{W}$ is the random vector which is experimentally observed and where $\mathbf{V}$ is another random response vector whose components are quantities which are not experimentally observed. As previously, $\mathbf{U}$ is independent of t, but depends on $\{\mathbf{Y}(t), t \in \mathcal{J}\}$, in which $\mathcal{J}$ is any part of $\mathbb{R}$ and where $\mathbf{Y}$ is the vector-valued stochastic process which is the second-order random solution of Eqs. (5.2) and (5.3). The probability density function $\mathbf{u} \mapsto p_{\mathbf{U}^{\text{post}}}(\mathbf{u})$ of the posterior random response vector $\mathbf{U}^{\text{post}}$ is then given by

$$p_{\mathbf{U}^{\text{post}}}(\mathbf{u}) = \int_{\mathbb{R}^{n_p}} p_{\mathbf{U}|\mathbf{X}}(\mathbf{u}|\mathbf{x})\, p_{\mathbf{X}}^{\text{post}}(\mathbf{x})\, d\mathbf{x}, \tag{5.11}$$

in which the conditional probability density function $p_{\mathbf{U}|\mathbf{X}}(\mathbf{u}|\mathbf{x})$ of $\mathbf{U}$, given $\mathbf{X} = \mathbf{x}$, is constructed using the stochastic computational model defined by Eqs. (5.2) and (5.3) with $\mathbf{X} = \mathbf{x}$. From Eq. (5.8), it can be deduced that

$$p_{\mathbf{U}^{\text{post}}}(\mathbf{u}) = E\{\mathcal{L}(\mathbf{X}^{\text{prior}})\, p_{\mathbf{U}|\mathbf{X}}(\mathbf{u}|\mathbf{X}^{\text{prior}})\}.$$

Let U_k^{post} be any component of random vector $\mathbf{U}^{\text{post}}$. The probability density function $u_k \mapsto p_{U_k^{\text{post}}}(u_k)$ on $\mathbb{R}$ of the posterior random variable U_k^{post} is then given by

$$p_{U_k^{\text{post}}}(u_k) = E\{\mathcal{L}(\mathbf{X}^{\text{prior}})\, p_{U_k|\mathbf{X}}(u_k|\mathbf{X}^{\text{prior}})\}, \tag{5.12}$$

in which $u_k \mapsto p_{U_k|\mathbf{X}}(u_k|\mathbf{x})$ is the conditional probability density function of the real valued random variable U_k given $\mathbf{X} = \mathbf{x}$ and which is also constructed using the stochastic computational model defined by Eqs. (5.2) and (5.3) with $\mathbf{X} = \mathbf{x}$.

5.4.3 Computational aspects

We use the notation introduced in Sections 5.1 and 5.2 concerning the realizations. Let $\mathbf{X}^{\text{prior}}(\theta_1), \ldots, \mathbf{X}^{\text{prior}}(\theta_\nu)$ be ν independent realizations of $\mathbf{X}^{\text{prior}}$ whose probability density function is $\mathbf{x} \mapsto p_{\mathbf{X}}^{\text{prior}}(\mathbf{x})$. For ν sufficiently large, the right-hand side of Eq. (5.12) can be estimated by

$$p_{U_k^{\text{post}}}(u_k) \simeq \frac{1}{\nu} \sum_{\ell=1}^{\nu} \mathcal{L}(\mathbf{X}^{\text{prior}}(\theta_\ell))\, p_{U_k|\mathbf{X}}(u_k|\mathbf{X}^{\text{prior}}(\theta_\ell)). \tag{5.13}$$

Let

$$([\mathbf{G}_M(\theta'_1)], [\mathbf{G}_D(\theta'_1)], [\mathbf{G}_K(\theta'_1)]), \ldots, [\mathbf{G}_M(\theta'_{\nu'})], [\mathbf{G}_D(\theta'_{\nu'})], [\mathbf{G}_K(\theta'_{\nu'})])$$

be ν' independent realizations of $([\mathbf{G}_M], [\mathbf{G}_D], [\mathbf{G}_K])$ whose probability density function is $([G_M], [G_D], [G_K]) \mapsto p_{\mathbf{G}}([G_M], [G_D], [G_K]; \boldsymbol{\delta}_G^{\text{opt}})$. For fixed θ_ℓ, the computational model defined by Eqs. (5.4) and (5.5) with $\mathbf{X}(\theta_\ell) = \mathbf{X}^{\text{prior}}(\theta_\ell) = \mathbf{x}$ is used to calculate the ν' independent realizations $\mathbf{U}(\theta'_1|\mathbf{x}), \ldots, \mathbf{U}(\theta'_{\nu'}|\mathbf{x})$. It can then be deduced $\mathbf{W}(\theta'_1|\mathbf{X}^{\text{prior}}(\theta_\ell)), \ldots,$ $\mathbf{W}(\theta'_{\nu'}|\mathbf{X}^{\text{prior}}(\theta_\ell))$ and, for all fixed k, we can compute $U_k(\theta'_1|\mathbf{X}^{\text{prior}}(\theta_\ell)),$ $\ldots, U_k(\theta'_{\nu'}|\mathbf{X}^{\text{prior}}(\theta_\ell))$.

(1) Using independent realizations $\mathbf{W}(\theta'_1|\mathbf{X}^{\text{prior}}(\theta_\ell)), \ldots, \mathbf{W}(\theta'_{\nu'}|\mathbf{X}^{\text{prior}}(\theta_\ell))$ and the multivariate Gaussian kernel density estimation method (see Section 8.6.5), for $r = 1, \ldots, \nu_{\text{exp}}$, we can estimate $p_{\mathbf{W}|\mathbf{X}}(\mathbf{w}_r^{\text{exp}}|\mathbf{X}^{\text{prior}}(\theta_\ell))$ and then, for $\ell = 1, \ldots, \nu$, we can deduce $\mathcal{L}(\mathbf{X}^{\text{prior}}(\theta_\ell))$ using Eq. (5.9).

(2) For all fixed k, $p_{U_k|\mathbf{X}}(u_k|\mathbf{X}^{\text{prior}}(\theta_\ell))$ is estimated using the independent realizations $U_k(\theta'_1|\mathbf{X}^{\text{prior}}(\theta_\ell)), \ldots, U_k(\theta'_{\nu'}|\mathbf{X}^{\text{prior}}(\theta_\ell))$ and the kernel estimation method (Bowman and Azzalini 1997). Using Eq. (5.13), it can then be deduced $p_{U_k^{\text{post}}}(u_k)$.

Chapter 6

Nonparametric probabilistic approach to uncertainties in structural-acoustic models for the low- and medium-frequency ranges

This section is devoted to the predictions of complex structural-acoustic systems in the low- and medium-frequency ranges for which computational structural-acoustic models are required. The presentation is limited to a bounded structure coupled with a bounded internal acoustic cavity. In order to simplify the presentation, the acoustic coupling of the structure with the unbounded external acoustic fluid is not considered here but can be taken into account without any difficulties (Ohayon and Soize 1998). For complex structural-acoustic systems, the mean computational structural-acoustic model is not sufficient to predict the responses of the real system due to the presence of both the model-parameter uncertainties and the model uncertainties induced by modeling errors in the computational model and by the variability of the real system with respect to the design system. In this section, we show how the nonparametric probabilistic approach of uncertainties presented in Section 4 can be used to model the uncertainties in the structural-acoustic computational model. We could present such an approach in the context of the generalized probabilistic approach presented in Section 5. However, we will limit the presentation to the context of the nonparametric probabilistic approach in order to simplify the presentation. Such an approach has been used in (Durand et al. 2008, Fernandez et al. 2009; 2010, Kassem et al. 2009; 2011) for structural-acoustic modeling of automotive vehicles in presence of uncertainties. The extension to the generalized probabilistic approach is straightforward and can easily be constructed in following the methodology introduced in Section 5.

6.1 Reduced mean structural-acoustic model

The structural-acoustic model is presented in the context of the 3D linear elastoacoustics for a structural-acoustic system made up of a damped elastic structure coupled with a closed internal acoustic cavity filled with a dissipative acoustic fluid. The linear responses of the structural-acoustic system are studied around a static equilibrium state which is taken as natural state at rest. The external acoustic fluid is a gas and its effects on the structural-acoustic system are assumed to be negligible in the frequency band of analysis. The mean computational structural-acoustic system is derived from the mean boundary value problem using the finite element method. Finally, the last step consists in constructing the reduced mean computational structural-acoustic model using modal analysis (Ohayon and Soize 1998).

6.1.1 Mean boundary value problem

Let Ω_s be a fixed damped structure subjected to external loads. The structure is coupled with its internal cavity Ω_a filled with a dissipative acoustic fluid. The frequency band of analysis is denoted by $\mathcal{F} = [\omega_{min}\,, \omega_{max}]$ rad/s. The three-dimensional space is referred to a cartesian system and its generic point is denoted by $\mathbf{x} = (x_1, x_2, x_3)$. The system is defined in Fig. 6.1. Let $\partial\Omega_s = \Gamma_s \cup \Gamma_0 \cup \Gamma_a$ be the boundary of Ω_s. The outward unit normal to $\partial\Omega_s$ is denoted by $\mathbf{n}_s = (n_{s,1}, n_{s,2}, n_{s,3})$. The displacement

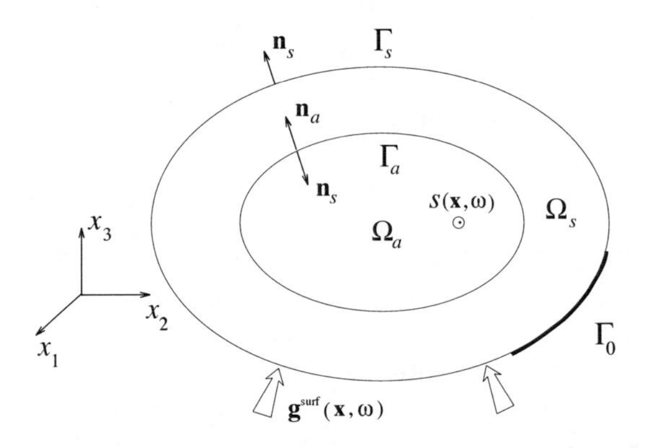

Figure 6.1. Scheme of the structural-acoustic system.

field in Ω_s is denoted by $\mathbf{u}(\mathbf{x},\omega) = (u_1(\mathbf{x},\omega), u_2(\mathbf{x},\omega), u_3(\mathbf{x},\omega))$. The structure is assumed to be fixed on the part Γ_0 of the boundary $\partial\Omega_s$. The internal acoustic cavity Ω_a is a bounded domain filled with a dissipative acoustic fluid. The boundary $\partial\Omega_a$ of Ω_a is Γ_a. The outward unit normal to $\partial\Omega_a$ is $\mathbf{n}_a = (n_{a,1}, n_{a,2}, n_{a,3})$ such that $\mathbf{n}_a = -\mathbf{n}_s$ on $\partial\Omega_a$. The pressure field in Ω_a is denoted by $p(\mathbf{x},\omega)$. For all ω in $\mathcal{F}$, the equation for the structure is written as

$$-\omega^2 \rho_s u_i - \frac{\partial \sigma_{ij}}{\partial x_j} = g_i^{vol} \quad \text{in} \quad \Omega_s, \tag{6.1}$$

with the convention for summations over repeated Latin indices, in which σ_{ij} is the stress tensor, $\mathbf{g}^{vol}(\mathbf{x},\omega) = (g_1^{vol}(\mathbf{x},\omega), g_2^{vol}(\mathbf{x},\omega), g_3^{vol}(\mathbf{x},\omega))$ is the body force field applied to the structure and ρ_s is the mass density. For $i = 1, 2$ or 3, the boundary conditions are

$$\sigma_{ij}(\mathbf{u}) n_{s,j} = g_i^{sur} \quad \text{on} \quad \Gamma_s, \quad \sigma_{ij}(\mathbf{u}) n_{s,j} = -p\, n_{s,i} \quad \text{on} \quad \Gamma_a, \tag{6.2}$$

$$u_i = 0 \quad \text{on} \quad \Gamma_0, \tag{6.3}$$

in which $\mathbf{g}^{sur}(\mathbf{x},\omega) = (g_1^{sur}(\mathbf{x},\omega), g_2^{sur}(\mathbf{x},\omega), g_3^{sur}(\mathbf{x},\omega))$ is the surface force field applied to the boundary Γ_s of the structure. The damped structure is made up of a linear non homogeneous anisotropic viscoelastic material. The constitutive equation is then written as $\sigma_{ij} = \mathbb{C}_{ijkh}(\mathbf{x})\epsilon_{kh}(\mathbf{u}) + i\omega\, \widetilde{\mathbb{C}}_{ijkh}(\mathbf{x},\omega)\varepsilon_{kh}(\mathbf{u})$. The tensor $\mathbb{C}_{ijkh}(\mathbf{x})$ of the elastic coefficients depends on $\mathbf{x}$ but is approximated by a function independent of ω for all ω in the frequency band of analysis $\mathcal{F}$. The tensor $\widetilde{\mathbb{C}}_{ijkh}(\mathbf{x},\omega)$ of the damping coefficients of the material depends on $\mathbf{x}$ and ω. These two tensors have the usual properties of symmetry and positive definiteness. The strain tensor $\varepsilon_{kh}(\mathbf{u})$ is related to the displacement field $\mathbf{u}$ by $\varepsilon_{kh}(\mathbf{u}) = (\partial u_k/\partial x_h + \partial u_h/\partial x_k)/2$. Concerning the internal dissipative acoustic fluid, the formulation in pressure is used. The equation governing the vibration of the dissipative acoustic fluid occupying domain Ω_a is written (Ohayon and Soize 1998) as

$$-\frac{\omega^2}{\rho_a c_a^2} p - i\omega \frac{\tau}{\rho_a} \nabla^2 p - \frac{1}{\rho_a} \nabla^2 p = -\frac{\tau}{\rho_a} c_a^2 \nabla^2 s + i \frac{\omega}{\rho_a} s \quad \text{in} \quad \Omega_a, \tag{6.4}$$

for which the boundary conditions are

$$\frac{1}{\rho_a}(1 + i\omega\tau) \frac{\partial p}{\partial \mathbf{n}_a} = \omega^2 \mathbf{u} \cdot \mathbf{n}_a + \tau \frac{c_a^2}{\rho_a} \frac{\partial s}{\partial \mathbf{n}_a} \quad \text{on} \quad \Gamma_a, \tag{6.5}$$

where ρ_a is the mass density of the acoustic fluid at equilibrium, c_a is the speed of sound, τ is the coefficient due to the viscosity of the fluid which can depends on ω (the coefficient due to thermal conduction is neglected) and where $s(\mathbf{x},\omega)$ is the acoustic source density.

6.1.2 Mean computational structural-acoustic model

The finite element method is used to approximate the boundary value problem defined by Eqs. (6.1) to (6.5). A finite element mesh of the structure Ω_s and the internal acoustic fluid Ω_a is then introduced. Let $\mathbb{u}(\omega) = (\mathbb{u}_1(\omega), \ldots, \mathbb{u}_{m_s}(\omega))$ be the complex vector of the m_s degrees-of-freedom of the structure according to the finite element discretization of the displacement field $\mathbf{x} \mapsto \mathbf{u}(\mathbf{x}, \omega)$. Let $\mathbb{p}(\omega) = (\mathbb{p}_1(\omega), \ldots, \mathbb{p}_{m_a}(\omega))$ be the complex vector of the m_a degrees-of-freedom of the acoustic fluid according to the finite element discretization of the pressure field $\mathbf{x} \mapsto p(\mathbf{x}, \omega)$. The finite element discretization of the boundary value problem yields the following mean computational structural-acoustic model,

$$\begin{bmatrix} [\mathbb{A}^s(\omega)] & [\mathbb{H}] \\ \omega^2 [\mathbb{H}]^T & [\mathbb{A}^a(\omega)] \end{bmatrix} \begin{bmatrix} \mathbb{u}(\omega) \\ \mathbb{p}(\omega) \end{bmatrix} = \begin{bmatrix} \mathbb{f}^s(\omega) \\ \mathbb{f}^a(\omega) \end{bmatrix}, \tag{6.6}$$

where $[\mathbb{A}^s(\omega)]$ is the dynamical stiffness matrix of the damped structure *in vacuo* which is a symmetric $(m_s \times m_s)$ complex matrix such that

$$[\mathbb{A}^s(\omega)] = -\omega^2 [\mathbb{M}^s] + i\omega [\mathbb{D}^s(\omega)] + [\mathbb{K}^s], \tag{6.7}$$

in which $[\mathbb{M}^s]$, $[\mathbb{D}^s(\omega)]$ and $[\mathbb{K}^s]$ are the mass, damping and stiffness matrices of the structure which belong to the set $\mathbb{M}^+_{m_s}(\mathbb{R})$ of all the positive-definite symmetric $(m_s \times m_s)$ real matrices. In Eq. (6.6), $[\mathbb{A}^a(\omega)]$ is the dynamical stiffness matrix of the dissipative acoustic fluid which is a symmetric $(m_a \times m_a)$ complex matrix such that

$$[\mathbb{A}^a(\omega)] = -\omega^2 [\mathbb{M}^a] + i\omega [\mathbb{D}^a(\omega)] + [\mathbb{K}^a], \tag{6.8}$$

in which $[\mathbb{M}^a]$, $[\mathbb{D}^a(\omega)]$ and $[\mathbb{K}^a]$ are the "mass", "damping" and "stiffness" matrices of the acoustic cavity with fixed coupling interface. The matrix $[\mathbb{M}^a]$ belongs to $\mathbb{M}^+_{m_a}(\mathbb{R})$ and, the matrices $[\mathbb{D}^a(\omega)]$ and $[\mathbb{K}^a]$ belong to the set $\mathbb{M}^{+0}_{m_a}(\mathbb{R})$ of all the semi-positive definite symmetric $(m_a \times m_a)$ real matrices for which the rank is $m_a - 1$. The matrix $[\mathbb{H}]$ is the structural-acoustic coupling matrix which belongs to the set $\mathbb{M}_{m_s, m_a}(\mathbb{R})$ of all the $(m_s \times m_a)$ real matrices.

6.1.3 Reduced mean computational structural-acoustic model

The projection of the mean computational structural-acoustic equation (6.6) on the structural modes *in vacuo* and on the acoustic modes of the acoustic cavity with fixed coupling interface yields the reduced mean computational structural-acoustic model (Ohayon and Soize 1998). The structural modes *in vacuo* and the acoustic modes of the cavity with fixed coupling interface are calculated by solving the two generalized eigenvalue problems

$$[\mathbb{K}^s]\,\boldsymbol{\phi}^s = \lambda^s[\mathbb{M}^s]\,\boldsymbol{\phi}^s, \quad [\mathbb{K}^a]\,\boldsymbol{\phi}^a = \lambda^a[\mathbb{M}^a]\,\boldsymbol{\phi}^a. \tag{6.9}$$

Let $[\Phi^s]$ be the $(m_s \times n)$ real matrix whose columns are the n structural modes $\boldsymbol{\phi}^s$ associated with the n first positive eigenvalues $0 < \lambda_1^s \leq \lambda_2^s \leq \ldots \leq \lambda_n^s$ (the n first structural eigenfrequencies are $\omega_\alpha^s = \sqrt{\lambda_\alpha^s}$). Let $[\Phi^a]$ be the $(m_a \times m)$ real matrix whose columns are constituted (1) of the constant pressure mode associated with the zero eigenvalue $\lambda_1 = 0$ and (2) of the acoustic modes $\boldsymbol{\phi}^a$ associated with the positive eigenvalues $0 < \lambda_2^a \leq \lambda_3^a \leq \ldots \leq \lambda_m^a$ (the $m-1$ first acoustical eigenfrequencies are $\omega_\beta^a = \sqrt{\lambda_\beta^a}$). It should be noted that the constant pressure mode is kept in order to model the quasi-static variation of the internal fluid pressure induced by the deformation of the coupling interface (Ohayon and Soize 1998). The eigenvectors satisfied the orthogonality conditions

$$[\Phi^s]^T[\mathbb{M}^s]\,[\Phi^s] = [I_n], \quad [\Phi^s]^T[\mathbb{K}^s]\,[\Phi^s] = [\lambda^s], \tag{6.10}$$

$$[\Phi^a]^T[\mathbb{M}^a]\,[\Phi^a] = [I_m], \quad [\Phi^a]^T[\mathbb{K}^a]\,[\Phi^a] = [\lambda^a], \tag{6.11}$$

in which $[\lambda^s]_{\alpha\alpha'} = \lambda_\alpha^s\,\delta_{\alpha\alpha'}$ and $[\lambda^a]_{\beta\beta'} = \lambda_\beta^a\,\delta_{\beta\beta'}$ are diagonal matrices, where $[I_n]$ and $[I_m]$ are the unity matrices of dimension n and m respectively, and where $\delta_{\alpha\alpha} = 1$ and $\delta_{\alpha\alpha'} = 0$ for $\alpha \neq \alpha'$. Using Eqs. (6.10) and (6.11), the projection of Eq. (6.6) yields the reduced mean matrix model of the structural-acoustic system (also called the reduced mean computational structural-acoustic model). For all ω in the frequency band $\mathcal{F}$, the structural displacement $\mathbb{u}^n(\omega)$ and the internal acoustic pressure $\mathbb{p}^m(\omega)$ are written as

$$\mathbb{u}^n(\omega) = [\Phi^s]\,\mathbf{q}^s(\omega), \quad \mathbb{p}^m(\omega) = [\Phi^a]\,\mathbf{q}^a(\omega), \tag{6.12}$$

in which the complex vectors $\mathbf{q}^s(\omega)$ and $\mathbf{q}^a(\omega)$ are such that

$$\begin{bmatrix} [A^s(\omega)] & [H] \\ \omega^2[H]^T & [A^a(\omega)] \end{bmatrix} \begin{bmatrix} \mathbf{q}^s(\omega) \\ \mathbf{q}^a(\omega) \end{bmatrix} = \begin{bmatrix} \mathbf{f}^s(\omega) \\ \mathbf{f}^a(\omega) \end{bmatrix}, \tag{6.13}$$

in which $[H] = [\Phi^s]^T\,[\mathbb{H}]\,[\Phi^a]$ is in $\mathbb{M}_{n,m}(\mathbb{R})$ and where $\mathbf{f}^s(\omega) = [\Phi^s]^T\,\mathbb{f}^s(\omega)$ and $\mathbf{f}^a(\omega) = [\Phi^a]^T\,\mathbb{f}^a(\omega)$ are complex vectors. The generalized dynamical stiffness matrix $[A^s(\omega)]$ of the damped structure is the $(n \times n)$ complex matrix written as

$$[A^s(\omega)] = -\omega^2[M^s] + i\omega[D^s(\omega)] + [K^s], \tag{6.14}$$

in which $[M^s] = [\Phi^s]^T [\mathbb{M}^s] [\Phi^s] = [I_n]$ and $[K^s] = [\Phi^s]^T [\mathbb{K}^s] [\Phi^s] = [\lambda^s]$ are diagonal matrices belonging to $\mathbb{M}_n^+(\mathbb{R})$. The generalized damping matrix $[D^s(\omega)] = [\Phi^s]^T [\mathbb{D}^s(\omega)] [\Phi^s]$ is a full matrix belonging to $\mathbb{M}_n^+(\mathbb{R})$. The generalized dynamical stiffness matrix $[A^a(\omega)]$ of the dissipative acoustic fluid is the $(m \times m)$ complex matrix written as

$$[A^a(\omega)] = -\omega^2[M^a] + i\omega[D^a(\omega)] + [K^a], \tag{6.15}$$

in which $[M^a] = [\Phi^a]^T [\mathbb{M}^a] [\Phi^a] = [I_m]$ is a diagonal matrix belonging to $\mathbb{M}_m^+(\mathbb{R})$ and where $[D^a(\omega)] = [\Phi^a]^T [\mathbb{D}^a(\omega)] [\Phi^a] = \tau(\omega) [\lambda^a]$ and $[K^a] = [\Phi^a]^T [\mathbb{K}^a] [\Phi^a] = [\lambda^a]$ are diagonal matrices belonging to $\mathbb{M}_m^{+0}(\mathbb{R})$ of rank $m - 1$.

Convergence. Below, we will denote by n_0 and m_0 the values of n and m for which the responses $\mathbb{u}^n$ and $\mathbb{p}^m$ are converged to $\mathbb{u}$ and $\mathbb{p}$, with a given accuracy.

6.2 Stochastic reduced-order model of the computational structural-acoustic model using the nonparametric probabilistic approach of uncertainties

For n and m fixed to the values n_0 and m_0 which have been previously defined, we use the nonparametric probabilistic approach of both the model-parameter uncertainties and the model uncertainties following the methodology presented in Section 4. For all ω in $\mathcal{F}$, the random vector $\mathbf{U}^n(\omega)$ of the structural displacements and the random vector $\mathbf{P}^m(\omega)$ of the acoustic pressures are given by

$$\mathbf{U}^n(\omega) = [\Phi^s]\, \mathbf{Q}^s(\omega), \quad \mathbf{P}^m(\omega) = [\Phi^a]\, \mathbf{Q}^a(\omega), \tag{6.16}$$

in which the random vectors $\mathbf{Q}^s(\omega)$ and $\mathbf{Q}^a(\omega)$ are such that

$$\begin{bmatrix} [\mathbf{A}^s(\omega)] & [\mathbf{H}] \\ \omega^2[\mathbf{H}]^T & [\mathbf{A}^a(\omega)] \end{bmatrix} \begin{bmatrix} \mathbf{Q}^s(\omega) \\ \mathbf{Q}^a(\omega) \end{bmatrix} = \begin{bmatrix} \mathbf{f}^s(\omega) \\ \mathbf{f}^a(\omega) \end{bmatrix}, \tag{6.17}$$

where $[\mathbf{H}]$ is a random matrix with values in the set $\mathbb{M}_{n,m}(\mathbb{R})$. The random generalized dynamical stiffness matrix of the damped structure is written as

$$[\mathbf{A}^s(\omega)] = -\omega^2[\mathbf{M}^s] + i\omega[\mathbf{D}^s(\omega)] + [\mathbf{K}^s], \tag{6.18}$$

in which $[\mathbf{M}^s]$, $[\mathbf{D}^s(\omega)]$ and $[\mathbf{K}^s]$ are random matrices with values in $\mathbb{M}_n^+(\mathbb{R})$. The random generalized dynamical stiffness matrix of the dissipative acoustic fluid is written as

$$[\mathbf{A}^a(\omega)] = -\omega^2[\mathbf{M}^a] + i\omega[\mathbf{D}^a(\omega)] + [\mathbf{K}^a], \tag{6.19}$$

in which $[\mathbf{M}^a]$ is a random matrix with values in $\mathbb{M}_m^+(\mathbb{R})$ and where $[\mathbf{D}^a(\omega)]$ and $[\mathbf{K}^a]$ are random matrices with values in $\mathbb{M}_m^{+0}(\mathbb{R})$. The rank of the two random matrices $[\mathbf{D}^a(\omega)]$ and $[\mathbf{K}^a]$ is fixed and is equal to $m-1$.

The development of the nonparametric probabilistic approach then requires to precisely define the available information for the random matrices above in order to construct their probability distributions using the maximum entropy principle which allows to maximize uncertainties. For all ω fixed in $\mathcal{F}$, this available information is defined as follows:

(i) The random matrices $[\mathbf{H}]$, $[\mathbf{M}^s]$, $[\mathbf{D}^s(\omega)]$, $[\mathbf{K}^s]$, $[\mathbf{M}^a]$, $[\mathbf{D}^a(\omega)]$ and $[\mathbf{K}^a]$ are defined on a probability space $(\Theta', \mathcal{T}', \mathcal{P}')$, and for θ' in Θ', we have

$$[\mathbf{H}(\theta')] \ \text{in} \ \mathbb{M}_{n,m}(\mathbb{R}), \tag{6.20}$$

$$[\mathbf{M}^s(\theta')]\,,\ [\mathbf{D}^s(\omega,\theta')] \ \text{and} \ [\mathbf{K}^s(\theta')] \ \text{in} \ \mathbb{M}_n^+(\mathbb{R}), \tag{6.21}$$

$$[\mathbf{M}^a(\theta')] \ \text{in} \ \mathbb{M}_m^+(\mathbb{R}) \quad ; \quad [\mathbf{D}^a(\omega,\theta')] \ \text{and} \ [\mathbf{K}^a(\theta')] \ \text{in} \ \mathbb{M}_m^{+0}(\mathbb{R}). \tag{6.22}$$

(ii) The mean value of these random matrices must be equal to the matrices of the reduced mean computational structural-acoustic model defined in Section 6.1.3:

$$E\{[\mathbf{H}]\} = [H], \tag{6.23}$$

$$E\{[\mathbf{M}^s]\} = [M^s]\,,\ E\{[\mathbf{D}^s(\omega)]\} = [D^s(\omega)]\,,\ E\{[\mathbf{K}^s]\} = [K^s], \tag{6.24}$$

$$E\{[\mathbf{M}^a]\} = [M^a]\,,\ E\{[\mathbf{D}^a(\omega)]\} = [D^a(\omega)]\,,\ E\{[\mathbf{K}^a]\} = [K^a], \tag{6.25}$$

in which $E\{[\mathbf{B}]\}$ denotes the mathematical expectation of random matrix $[\mathbf{B}]$.

(iii) For all ω fixed in $\mathcal{F}$, the probability distributions of the random matrices $[\mathbf{H}]$, $[\mathbf{M}^s]$, $[\mathbf{D}^s(\omega)]$, $[\mathbf{K}^s]$, $[\mathbf{M}^a]$, $[\mathbf{D}^a(\omega)]$ and $[\mathbf{K}^a]$ must be such that the random solution $(\mathbf{U}^n(\omega), \mathbf{P}^m(\omega))$ of Eqs. (6.16) to (6.17) is a second-order random vector,

$$E\{\|\mathbf{U}^n(\omega)\|^2\} < +\infty, \quad E\{\|\mathbf{P}^m(\omega)\|^2\} < +\infty. \tag{6.26}$$

6.3 Construction of the prior probability model of uncertainties

(i) Let us assumed that $n = n_0$ and $m = m_0$ in which n_0 and m_0 are defined in Section 6.1.3, with $n \geq m$. The elastic modes of the structure and the acoustic modes of the internal acoustic cavity are selected in order for that the matrix $[H]$ which belongs to $\mathbb{M}_{n,m}(\mathbb{R})$ be such that $[H]\,\mathbf{q}^a = 0$ implies $\mathbf{q}^a = 0$. It should be noted that if $m \geq n$ the following construction must be applied to $[H]^T$ instead of $[H]$. Then, using the singular value decomposition of the rectangular matrix $[H]$, one can write $[H] = [C]\,[T]$ in which the rectangular matrix $[C]$ in $\mathbb{M}_{n,m}(\mathbb{R})$ is such that $[C]^T\,[C] = [\,I_m]$ and where the symmetric square matrix $[T]$ belongs to $\mathbb{M}_m^+(\mathbb{R})$ and can then be written (using the Cholesky decomposition) as $[T] = [L_T]^T\,[L_T]$ in which $[L_T]$ is an upper triangular matrix in $\mathbb{M}_m(\mathbb{R})$. Since $[T]$ belongs to $\mathbb{M}_m^+(\mathbb{R})$, we introduce the random matrix $[\mathbf{T}]$ with values in $\mathbb{M}_m^+(\mathbb{R})$ such that $E\{[\mathbf{T}]\} = [T]$. The random matrix $[\mathbf{H}]$ with values in $\mathbb{M}_{n,m}(\mathbb{R})$ can then be written (Soize 2005b) as

$$[\mathbf{H}] = [C]\,[\mathbf{T}], \quad [\mathbf{T}] = [L_T]^T\,[\mathbf{G}_H]\,[L_T], \tag{6.27}$$

with $[\mathbf{G}_H]$ a random matrix with values in $\mathbb{M}_m^+(\mathbb{R})$ such that $E\{[\mathbf{G}_H]\} = [I_m]$.

(ii) For all ω in $\mathcal{F}$, it can be seen that the positive-definite matrices $[M^s]$, $[K^s]$, $[D^s(\omega)]$, $M^a]$ and the semi-positive definite matrices $[K^a]$, $[D^a(\omega)]$ can be written as

$$[M^s] = [L_{M^s}]^T\,[L_{M^s}], \quad [K^s] = [L_{K^s}]^T\,[L_{K^s}], \tag{6.28}$$

$$[D^s(\omega)] = [L_{D^s}(\omega)]^T\,[L_{D^s}(\omega)], \quad [M^a] = [L_{M^a}]^T\,[L_{M^a}], \tag{6.29}$$

$$[K^a] = [L_{K^a}]^T\,[L_{K^a}], \quad [D^a(\omega)] = [L_{D^a}(\omega)]^T\,[L_{D^a}(\omega)], \tag{6.30}$$

in which $[L_{M^s}]$ is the unity matrix $[I_n]$, $[L_{K^s}]$ is the diagonal matrix $[\lambda^s]^{1/2}$ in $\mathbb{M}_n(\mathbb{R})$, $[L_{D^s}(\omega)]$ is an upper triangular matrix in $\mathbb{M}_n(\mathbb{R})$, $[L_{M^a}]$ is the unity matrix $[I_m]$, $[L_{K^a}]$ is a rectangular matrix in $\mathbb{M}_{m-1,m}(\mathbb{R})$ such that, for all $\beta = 1, \ldots, m-1$, we have $[L_{K^a}]_{\beta 1} = 0$ and for $\beta' = 2, \ldots, m$, we have $[L_{K^a}]_{\beta\beta'} = \sqrt{\lambda^a_{\beta'}}\,\delta_{(\beta+1)\beta'}$. Finally, $[L_{D^a}(\omega)] = \sqrt{\tau(\omega)}\,[L_{K^a}]$ is a rectangular matrix in $\mathbb{M}_{m-1,m}(\mathbb{R})$. The random matrices $[\mathbf{M}^s]$, $[\mathbf{K}^s]$ and $[\mathbf{D}^s(\omega)]$ are written as

$$[\mathbf{M}^s] = [L_{M^s}]^T\,[\mathbf{G}_{M^s}]\,[L_{M^s}], \quad [\mathbf{K}^s] = [L_{K^s}]^T\,[\mathbf{G}_{K^s}]\,[L_{K^s}], \tag{6.31}$$

$$[\mathbf{D}^s(\omega)] = [L_{D^s}(\omega)]^T\,[\mathbf{G}_{D^s}]\,[L_{D^s}(\omega)], \tag{6.32}$$

in which $[\mathbf{G}_{M^s}]$, $[\mathbf{G}_{K^s}]$ and $[\mathbf{G}_{D^s}]$ are random matrices with values in $\mathbb{M}_n^+(\mathbb{R})$. These three random matrices are independent of ω and such that $E\{[\mathbf{G}_{M^s}]\} = E\{[\mathbf{G}_{K^s}]\} = E\{[\mathbf{G}_{D^s}]\} = [I_n]$. Similarly, the random matrices $[\mathbf{M}^a]$, $[\mathbf{K}^a]$ and $[\mathbf{D}^a(\omega)]$ are written as

$$[\mathbf{M}^a] = [L_{M^a}]^T\,[\mathbf{G}_{M^a}]\,[L_{M^a}], \quad [\mathbf{K}^a] = [L_{K^a}]^T\,[\mathbf{G}_{K^a}]\,[L_{K^a}], \tag{6.33}$$

$$[\mathbf{D}^a(\omega)] = [L_{D^a}(\omega)]^T\,[\mathbf{G}_{D^a}]\,[L_{D^a}(\omega)], \tag{6.34}$$

in which $[\mathbf{G}_{M^a}]$ is a random matrix with values in $\mathbb{M}_m^+(\mathbb{R})$, independent of ω and such that $E\{[\mathbf{G}_{M^a}]\} = [I_m]$ and where $[\mathbf{G}_{K^a}]$ and $[\mathbf{G}_{D^a}]$ are random matrices with values in $\mathbb{M}_{m-1}^+(\mathbb{R})$, independent of ω and such that $E\{[\mathbf{G}_{K^a}]\} = E\{[\mathbf{G}_{D^a}]\} = [I_{m-1}]$.

(iii) Taking into account Eqs. (6.27) to (6.34), the stochastic reduced computational model is completely defined as soon as the joint probability distribution of the random matrices $[\mathbf{G}_H]$, $[\mathbf{G}_{M^s}]$, $[\mathbf{G}_{D^s}]$, $[G_{K^s}]$, $[\mathbf{G}_{M^a}]$, $[\mathbf{G}_{D^a}]$ and $[\mathbf{G}_{K^a}]$ are constructed. Let $[\mathbf{G}]$ be any one of these random matrices with values in $\mathbb{M}_\mu^+(\mathbb{R})$ for which μ is equal to n, m or $m-1$. We then have $E\{[\mathbf{G}]\} = [I_\mu]$ and it has been proven that $E\{\|[\mathbf{G}]^{-1}\|_F^2\}$ must be finite in order that Eq. (6.26) be verified. Consequently, random matrix $[\mathbf{G}]$ must belong to ensemble $\mathrm{SG}_\varepsilon^+$ defined in Section 2.5.3.

6.4 Model parameters, stochastic solver and convergence analysis

(i) In Section 2.5.3, we have seen that the dispersion parameter δ allowed the dispersion of the random matrix $[\mathbf{G}]$ to be controlled. The dispersion parameters of random matrices $[\mathbf{G}_H]$, $[\mathbf{G}_{M^s}]$, $[\mathbf{G}_{D^s}]$, $[\mathbf{G}_{K^s}]$, $[\mathbf{G}_{M^a}]$, $[\mathbf{G}_{D^a}]$ and $[\mathbf{G}_{K^a}]$ will then be denoted by

$$\delta_H\,,\ \delta_{M^s}\,,\ \delta_{D^s}\,,\ \delta_{K^s}\,,\ \delta_{M^a}\,,\ \delta_{D^a}\,,\ \delta_{K^a}\,. \tag{6.35}$$

and allow the level of uncertainties to be controlled.

(ii) For n and m fixed to the values n_0 and m_0, and for all ω fixed in $\mathcal{F}$, the random equation defined by Eq. (6.17) is solved by the Monte Carlo method with ν independent realizations. Using the usual statistical estimator of the mathematical expectation E, the mean-square convergence of the stochastic solution with respect to ν can be analyzed in studying the function $\nu \mapsto \mathcal{E}(\nu)$ such that

$$\mathcal{E}(\nu) = \nu^{-1/2}\{\sum_{\ell=1}^{\nu}\int_{\mathcal{F}}(\|\mathbf{U}^n(\omega,\theta_\ell)\|^2 + \|\mathbf{P}^m(\omega,\theta_\ell)\|^2)\,d\omega\}^{1/2}\,. \tag{6.36}$$

6.5 Estimation of the parameters of the prior probability model of uncertainties

There are several possible methodologies to identify the optimal values of the dispersion parameters δ_{M^s}, δ_{D^s}, δ_{K^s} for the structure, δ_{M^a}, δ_{D^a},

δ_{K^a} for the internal acoustic cavity and δ_H for the structural-acoustic coupling interface, using experimental data. The methodology described in Section 4.3 can be used. For the identification using experimental data, many details can be found in (Durand et al. 2008, Soize et al. 2008a, Fernandez et al. 2009; 2010).

6.6 Comments about the applications and the experimental validation of the nonparametric probabilistic approach of uncertainties in structural acoustics

Concerning the applications to structural-acoustic problems, we refer the reader to (Durand et al. 2008, Fernandez et al. 2009, Kassem et al. 2009, Fernandez et al. 2010, Kassem et al. 2011) for structural acoustics of automotive vehicles with experimental comparisons, and to (Capiez-Lernout and Soize 2008a) for an example of the design optimization with an uncertain structural-acoustic model.

Chapter 7

Nonparametric probabilistic approach to uncertainties in computational nonlinear structural dynamics

In Section 4, we have presented the nonparametric probabilistic approach of uncertainties in computational structural dynamics which allows both the model-parameter uncertainties and the model uncertainties induced by modeling errors to be taken into account. In Section 5, we have presented the generalized probabilistic approach of uncertainties which allows the model-parameter uncertainties and the model uncertainties to be separately modeled. In this section, we are interested in the nonparametric probabilistic approach of uncertainties in computational nonlinear structural dynamics (it should be noted that the extension to the generalized probabilistic approach is straightforward) . We then consider uncertain infinite-dimensional geometrically nonlinear dynamical systems for which uncertainties (both the uncertainties on the computational model parameters and the model uncertainties) are taken into account using the nonparametric probabilistic approach.

Below, we then summarize the methodology developed and validated in (Mignolet and Soize 2007; 2008b, Wang et al. 2010). The nonparametric probabilistic approach of uncertainties is used to analyze the robustness of the solutions for uncertain infinite-dimensional geometrically nonlinear dynamical systems. Such an approach is particularly computationally attractive as it applies to reduced-order computational models of the dynamical system. The nonparametric probabilistic approach of uncertainties, originally developed for linear operators and presented in Section 4, has been extended to the nonlinear operator of the 3D geometrically nonlinear elasticity and has required additional developments.

7.1 Nonlinear equation for 3D geometrically nonlinear elasticity

Let $\mathbf{x} = (x_1, x_2, x_3)$ be the generic point in $\mathbb{R}^3$. Let $\{\mathbf{u}(\mathbf{x}, t), \mathbf{x} \in \Omega_0\}$ be the vector-valued displacement field expressed in the undeformed configuration Ω_0 (bounded domain of $\mathbb{R}^3$), with boundary $\partial\Omega_0 = \partial\Omega_0^0 \cup \partial\Omega_0^1$. Let $\mathbf{n}^0 = (n_1^0, n_2^0, n_3^0)$ be the external unit normal to $\partial\Omega_0$. The boundary value problem related to the considered geometrically nonlinear dynamical system is written (Mignolet and Soize 2008b) as

$$
\begin{aligned}
\rho_0\, \ddot{u}_i - \tfrac{\partial}{\partial x_k}(F_{ij} S_{jk}) = \rho_0\, b_i^0 \quad \text{in} \quad \Omega_0, \\
\mathbf{u} = 0 \quad \text{on} \quad \partial\Omega_0^0, \\
F_{ij} S_{jk} n_k^0 = g_i^0 \quad \text{on} \quad \partial\Omega_0^1, \\
S_{jk} = \mathbb{C}_{jk\ell m}\, E_{\ell m} \quad ,
\end{aligned}
\tag{7.1}
$$

with the convention for summations over repeated Latin indices, in which $\rho_0(\mathbf{x}) > 0$ is the density in the undeformed configuration, $\mathbf{b}^0(\mathbf{x}) = (b_1^0(\mathbf{x}), b_2^0(\mathbf{x}), b_3^0(\mathbf{x}))$ is the vector of body forces, $\mathbf{g}^0(\mathbf{x}) = (g_1^0(\mathbf{x}), g_2^0(\mathbf{x}), g_3^0(\mathbf{x}))$ is the vector of surface forces, $S_{jk} = \mathbb{C}_{jk\ell m}(\mathbf{x})\, E_{\ell m}$ is the constitutive equation for the linear elastic material with fourth-order elasticity tensor $\mathbb{C}_{jk\ell m}(\mathbf{x})$, where S_{jk} is the second Piola-Kirchhoff stress tensor and $E_{\ell m} = \frac{1}{2}(F_{k\ell}\, F_{km} - \delta_{\ell m})$ is the Green strain tensor with F_{ij} the deformation gradient tensor such that

$$
F_{ij} = \partial(u_i + x_i)/\partial x_j = \delta_{ij} + \partial u_i/\partial x_j . \tag{7.2}
$$

Let $\mathcal{C}_{\text{ad}}$ be the admissible space defined by

$$
\mathcal{C}_{\text{ad}} = \{\mathbf{v} \in \Omega_0 \ , \ \mathbf{v} \text{ sufficiently regular} \ , \ \mathbf{v} = 0 \text{ on } \partial\Omega_0^0\}. \tag{7.3}
$$

The weak formulation of the boundary value problem consists in finding the unknown displacement field $\mathbf{u}(., t)$ in $\mathcal{C}_{\text{ad}}$ such that, for all $\mathbf{v}$ in $\mathcal{C}_{\text{ad}}$,

$$
\int_{\Omega_0} \rho_0\, \ddot{u}_i v_i \, d\mathbf{x} + \int_{\Omega_0} F_{ij} S_{jk} \frac{\partial v_i}{\partial x_k} \, d\mathbf{x} = \int_{\Omega_0} \rho_0\, b_i^0\, v_i \, d\mathbf{x} + \int_{\partial\Omega_0^1} g_i^0\, v_i \, ds. \tag{7.4}
$$

7.2 Nonlinear reduced mean model

The vector-valued displacement field $\{\mathbf{u}(\mathbf{x}, t), \mathbf{x} \in \Omega_0\}$ is approximated by

$$
\mathbf{u}^n(\mathbf{x}, t) = \sum_{\alpha=1}^{n} q_\alpha(t)\, \boldsymbol{\varphi}^\alpha(\mathbf{x}), \tag{7.5}
$$

with n the dimension of the reduced-order model and where $\boldsymbol{\varphi}^1, \ldots, \boldsymbol{\varphi}^n$ is an adapted basis which can be constructed, for instance, using either the elastic modes of the linearized system, either the elastic modes of the underlying linear system or the POD modes of the nonlinear system (Azeez and Vakakis 2001, Holmes et al. 1997, Kunisch and Volkwein 2001; 2002, Sampaio and Soize 2007b). From the weak formulation of the boundary value problem, defined by Eq. (7.4), it can be deduced that the vector $\mathbf{q}(t) = (q_1(t), \ldots, q_n(t))$ of the generalized coordinates, verifies the following nonlinear reduced-order equation,

$$[M]\ddot{\mathbf{q}}(t) + [D]\dot{\mathbf{q}}(t) + \mathbf{k}_{\mathrm{NL}}(\mathbf{q}(t)) = \mathbf{f}(t), \tag{7.6}$$

in which $[M]$ is the generalized mass matrix, $[D]$ is the generalized damping matrix and $\mathbf{f}(t)$ is the generalized external force vector. The nonlinear function $\mathbf{q} \mapsto \mathbf{k}_{\mathrm{NL}}(\mathbf{q})$ is given by

$$\{\mathbf{k}_{\mathrm{NL}}(\mathbf{q})\}_\alpha = \sum_{\beta=1}^{n} K^{(1)}_{\alpha\beta}\, q_\beta + \sum_{\beta,\gamma=1}^{n} K^{(2)}_{\alpha\beta\gamma}\, q_\beta q_\gamma + \sum_{\beta,\gamma,\delta=1}^{n} K^{(3)}_{\alpha\beta\gamma\delta}\, q_\beta q_\gamma q_\delta . \tag{7.7}$$

It can then be proven that the second PiolaKirchhoff stress tensor can be written as

$$S_{jk}(t) = S^{(0)}_{jk} + \sum_{\alpha=1}^{n} S^{(1)\alpha}_{jk} q_\alpha(t) + \sum_{\alpha,\beta=1}^{n} S^{(2)\alpha,\beta}_{jk}\, q_\alpha(t)\, q_\beta(t). \tag{7.8}$$

The second-order tensor $K^{(1)}_{\alpha\beta}$, the third-order tensor $K^{(2)}_{\alpha\beta\gamma}$ and the fourth-order tensor $K^{(3)}_{\alpha\beta\gamma\delta}$ are given by

$$K^{(1)}_{\alpha\beta} = \int_{\Omega_0} \mathbb{C}_{jk\ell m}(\mathbf{x}) \frac{\partial \varphi^\alpha_j(\mathbf{x})}{\partial x_k} \frac{\partial \varphi^\beta_\ell(\mathbf{x})}{\partial x_m}\, d\mathbf{x}, \tag{7.9}$$

$$K^{(2)}_{\alpha\beta\gamma} = \frac{1}{2}\Big(\widehat{K}^{(2)}_{\alpha\beta\gamma} + \widehat{K}^{(2)}_{\beta\gamma\alpha} + \widehat{K}^{(2)}_{\gamma\alpha\beta}\Big), \tag{7.10}$$

$$\widehat{K}^{(2)}_{\alpha\beta\gamma} = \int_{\Omega_0} \mathbb{C}_{jk\ell m}(\mathbf{x}) \frac{\partial \varphi^\alpha_j(\mathbf{x})}{\partial x_k} \frac{\partial \varphi^\beta_s(\mathbf{x})}{\partial x_\ell} \frac{\partial \varphi^\gamma_s(\mathbf{x})}{\partial x_m}\, d\mathbf{x}, \tag{7.11}$$

$$K^{(3)}_{\alpha\beta\gamma\delta} = \int_{\Omega_0} \mathbb{C}_{jk\ell m}(\mathbf{x}) \frac{\partial \varphi^\alpha_r(\mathbf{x})}{\partial x_j} \frac{\partial \varphi^\beta_r(\mathbf{x})}{\partial x_k} \frac{\partial \varphi^\gamma_s(\mathbf{x})}{\partial x_\ell} \frac{\partial \varphi^\delta_s(\mathbf{x})}{\partial x_m}\, d\mathbf{x}. \tag{7.12}$$

It can easily be seen that the symmetry properties of the fourth-order elasticity tensor $\mathbb{C}_{jk\ell m}(\mathbf{x})$ yield the following properties

$$K^{(1)}_{\alpha\beta} = K^{(1)}_{\beta\alpha}, \tag{7.13}$$

$$\widehat{K}^{(2)}_{\alpha\beta\gamma} = \widehat{K}^{(2)}_{\alpha\gamma\beta}, \tag{7.14}$$

$$K^{(2)}_{\alpha\beta\gamma} = K^{(2)}_{\beta\gamma\alpha} = K^{(2)}_{\gamma\alpha\beta}, \tag{7.15}$$

$$K^{(3)}_{\alpha\beta\gamma\delta} = K^{(3)}_{\alpha\beta\delta\gamma} = K^{(3)}_{\beta\alpha\gamma\delta} = K^{(3)}_{\gamma\delta\alpha\beta}. \tag{7.16}$$

Moreover, using the positive-definite property of the fourth-order elasticity tensor $\mathbb{C}_{jk\ell m}(\mathbf{x})$, it can be shown that tensors $K^{(1)}_{\alpha\beta}$ and $K^{(3)}_{\alpha\beta\gamma\delta}$ are positive definite.

Convergence. Below, we will denote by n_0 the value of n for which the response $\mathbf{u}^n$ is converged to $\mathbf{u}$, with a given accuracy.

7.3 Algebraic properties of the nonlinear stiffnesses

Tensor $\widehat{K}^{(2)}_{\alpha\beta\gamma}$ is reshaped into a $n \times n^2$ matrix and tensor $K^{(3)}_{\alpha\beta\gamma\delta}$ is reshaped into a $n^2 \times n^2$ matrix such that

$$[\widehat{K}^{(2)}]_{\alpha J} = \widehat{K}^{(2)}_{\alpha\beta\gamma}, \quad J = n(\beta - 1) + \gamma, \tag{7.17}$$

$$[K^{(3)}]_{IJ} = K^{(3)}_{\alpha\beta\gamma\delta}, \quad I = n(\alpha - 1) + \beta, \quad J = n(\gamma - 1) + \delta. \tag{7.18}$$

Let $[K_B]$ be the $m \times m$ matrix with $m = n + n^2$ such that

$$[K_B] = \begin{bmatrix} [K^{(1)}] & [\widehat{K}^{(2)}] \\ [\widehat{K}^{(2)}]^T & 2[K^{(3)}] \end{bmatrix}. \tag{7.19}$$

In (Mignolet and Soize 2008b), it is proven that the symmetric real matrix $[K_B]$ is positive definite. This strong property will be used to construct the nonparametric probabilistic approach of uncertainties.

7.4 Stochastic reduced-order model of the nonlinear dynamical system using the nonparametric probabilistic approach of uncertainties

For n fixed to the value n_0 defined at the end of Section 7.2, the use of the nonparametric probabilistic approach of uncertainties consists (Mignolet and Soize 2008b) in substituting the deterministic positive-definite matrices $[M]$, $[D]$ and $[K_B]$ by random matrices $[\mathbf{M}]$, $[\mathbf{D}]$ and $[\mathbf{K}_B]$ belonging to adapted ensembles of random matrices.

7.4.1 Construction of the probability distributions of the random matrices

The random matrices $[\mathbf{M}]$, $[\mathbf{D}]$ and $[\mathbf{K}_B]$ are then written as

$$[\mathbf{M}] = [L_M]^T \, [\mathbf{G}_M] \, [L_M], \quad [\mathbf{D}] = [L_D]^T \, [\mathbf{G}_D] \, [L_D], \tag{7.20}$$

$$[\mathbf{K}_B] = [L_{K_B}]^T \, [\mathbf{G}_{K_B}] \, [L_{K_B}], \tag{7.21}$$

in which $[L_M]$, $[L_D]$ and $[L_{K_B}]$ are the upper triangular matrices such that

$$[M] = [L_M]^T[L_M], \quad [D] = [L_D]^T[L_D], \quad [K_B] = [L_{K_B}]^T[L_{K_B}].$$

In Eqs. (7.20) and (7.21), the random matrices $[\mathbf{G}_M]$, $[\mathbf{G}_D]$ and $[\mathbf{G}_{K_B}]$ are random matrices which are defined on probability space $(\Theta', \mathcal{T}', \mathcal{P}')$, which are with values in $\mathbb{M}_n^+(\mathbb{R})$, $\mathbb{M}_n^+(\mathbb{R})$ and $\mathbb{M}_m^+(\mathbb{R})$ respectively, and which belong to the ensemble $\mathrm{SG}_\varepsilon^+$ of random matrices defined in Section 2.5.3. These three random matrices are then statistically independent and depend on the positive real dispersion parameters δ_M, δ_D and δ_{K_B} which allow the level of uncertainties to be controlled.

7.4.2 Stochastic reduced-order model of the nonlinear dynamical system

Taking into account Eqs. (7.19) and (7.21), the deterministic matrices $[K^{(1)}]$, $[\widehat{K}^{(2)}]$ and $[K^{(3)}]$ are then replaced by three dependent random matrices $[\mathbf{K}^{(1)}]$, $[\widehat{\mathbf{K}}^{(2)}]$ and $[\mathbf{K}^{(3)}]$ such that

$$\begin{bmatrix} [\mathbf{K}^{(1)}] & [\widehat{\mathbf{K}}^{(2)}] \\ [\widehat{\mathbf{K}}^{(2)}]^T & 2[\mathbf{K}^{(3)}] \end{bmatrix} = [\mathbf{K}_B]. \tag{7.22}$$

Let $\widehat{\mathbf{K}}^{(2)}_{\alpha\beta\gamma} = [\widehat{\mathbf{K}}^{(2)}]_{\alpha J}$ be the third-order random tensor corresponding to the random matrix $[\widehat{\mathbf{K}}^{(2)}]$ and obtained by the back reshaping process. Taking into account Eq. (7.10), the third-order deterministic tensor $K^{(2)}_{\alpha\beta\gamma}$ is then replaced by the third-order random tensor $\mathbf{K}^{(2)}_{\alpha\beta\gamma}$ such that

$$\mathbf{K}^{(2)}_{\alpha\beta\gamma} = \frac{1}{2}\left(\widehat{\mathbf{K}}^{(2)}_{\alpha\beta\gamma} + \widehat{\mathbf{K}}^{(2)}_{\beta\gamma\alpha} + \widehat{\mathbf{K}}^{(2)}_{\gamma\alpha\beta}\right). \tag{7.23}$$

Let $\mathbf{K}^{(3)}_{\alpha\beta\gamma\delta} = [\mathbf{K}^{(3)}]_{IJ}$ be the fourth-order random tensor corresponding to the random matrix $[\mathbf{K}^{(3)}]$ and obtained by the back reshaping process.

Taking into account Eqs. (7.5) to (7.7), the stochastic reduced-order model of the nonlinear dynamical system using the nonparametric probabilistic approach of uncertainties is written as

$$\mathbf{U}(\mathbf{x},t) = \sum_{\alpha=1}^{n} Q_\alpha(t)\, \boldsymbol{\varphi}^\alpha(\mathbf{x}), \tag{7.24}$$

in which the vector-valued stochastic process $\mathbf{Q}(t) = (Q_1(t),\ldots,Q_n(t))$ verifies the stochastic nonlinear equation

$$[\mathbf{M}]\,\ddot{\mathbf{Q}}(t) + [\mathbf{D}]\,\dot{\mathbf{Q}}(t) + \mathbf{K}_{\text{NL}}(\mathbf{Q}(t)) = \mathbf{f}(t). \tag{7.25}$$

The random nonlinear function $\mathbf{q} \mapsto \mathbf{K}_{\text{NL}}(\mathbf{q})$ is written as

$$\{\mathbf{K}_{\text{NL}}(\mathbf{q})\}_\alpha = \sum_{\beta=1}^{n} \mathbf{K}^{(1)}_{\alpha\beta}\, q_\beta + \sum_{\beta,\gamma=1}^{n} \mathbf{K}^{(2)}_{\alpha\beta\gamma}\, q_\beta q_\gamma + \sum_{\beta,\gamma,\delta=1}^{n} \mathbf{K}^{(3)}_{\alpha\beta\gamma\delta}\, q_\beta q_\gamma q_\delta\,. \tag{7.26}$$

7.5 Comments about the applications of the nonparametric probabilistic approach of uncertainties in computational nonlinear structural dynamics

Concerning the applications of the nonparametric probabilistic approach of uncertainties in computational nonlinear structural dynamics in the context of the 3D geometrically nonlinear elasticity, we refer the reader to (Kim et al. 2007, Mignolet and Soize 2007; 2008b, Wang et al. 2010).

Chapter 8

Identification of high-dimension polynomial chaos expansions with random coefficients for non-Gaussian tensor-valued random fields using partial and limited experimental data

In this section, we present a new methodology proposed in (Soize 2010b; 2011) to solve the challenging problem related to the identification of random Vector-Valued Coefficients (VVC) of the high-dimension Polynomial Chaos Expansions (PCE) of non-Gaussian tensor-valued random fields using partial and limited experimental data. The experimental data sets correspond to an observation vector which is the response of a stochastic boundary value problem depending on the tensor-valued random field which has to be identified. So an inverse stochastic problem must be solved to perform the identification of the random field. Such a problem is met, for instance, in mechanics of materials with heterogeneous complex microstructures.

8.1 Definition of the problem to be solved

(1) Stochastic boundary value problem

We consider a boundary value problem for a vector-valued field $\{\mathbf{u}(\mathbf{x}) = (u_1(\mathbf{x}), u_2(\mathbf{x}), u_3(\mathbf{x})), \mathbf{x} \in \Omega\}$ defined on an open bounded domain Ω of $\mathbb{R}^3$, with generic point $\mathbf{x} = (x_1, x_2, x_3)$. This boundary value problem depends on a non-Gaussian fourth-order tensor-valued random field $\{\mathbb{C}(\mathbf{x}), \mathbf{x} \in \Omega\}$ with $\mathbb{C}(\mathbf{x}) = \{\mathbb{C}_{ijk\ell}(\mathbf{x})\}_{ijk\ell}$, which is unknown and which has to be identified solving an inverse stochastic problem. The boundary $\partial\Omega$ of domain Ω is written as $\Gamma_0 \cup \Gamma_{\text{obs}} \cup \Gamma$. Field $\mathbf{u}$ is experimentally observed on Γ_{obs}, which means that the system is partially observed with respect to the available experimental data. It is assumed that there is a zero Dirichlet condition $\mathbf{u} = 0$ on Γ_0 (such an assumption could be released in the following).

(2) Stochastic finite element approximation of the stochastic boundary value problem

The weak formulation of the stochastic boundary value problem (introduced in (1) above) is discretized by the finite element method. Let $\mathcal{I} = \{\mathbf{x}^1, \ldots, \mathbf{x}^{N_p}\} \subset \Omega$ be the finite subset of Ω made up of all the integrations points of the finite elements used in the mesh of Ω. For all $\mathbf{x}$ fixed in $\mathcal{I} \subset \Omega$, the fourth-order tensor-valued random variable $\mathbb{C}(\mathbf{x})$ is represented by a real random matrix $[\mathbb{A}(\mathbf{x})]$ such that

$$[\mathbb{A}(\mathbf{x})]_{IJ} = \mathbb{C}_{ijkh}(\mathbf{x}) \quad \text{with} \quad I = (i, j) \quad \text{and} \quad J = (k, h). \tag{8.1}$$

It should be noted that mathematical properties on the matrix-valued random field $\{[\mathbb{A}(\mathbf{x})], \mathbf{x} \in \Omega\}$ are necessary in order to preserve the mathematical properties of the boundary value problem (see Section 8.2). Let $\mathbf{U} = (\mathbf{U}^{\text{obs}}, \mathbf{U}^{\text{nobs}})$ be the random vector with values in $\mathbb{R}^m = \mathbb{R}^{m_{\text{obs}}} \times \mathbb{R}^{m_{\text{nobs}}}$ with $m = m_{\text{obs}} + m_{\text{nobs}}$, constituted of some degrees of freedom of the finite element approximation of field $\mathbf{u}$. The $\mathbb{R}^{m_{\text{obs}}}$-valued random vector $\mathbf{U}^{\text{obs}} = (U_1^{\text{obs}}, \ldots, U_{m_{\text{obs}}}^{\text{obs}})$ is made up of the m_{obs} observed degrees of freedom for which there are available experimental data (corresponding to the finite element approximation of the trace on Γ_{obs} of random field $\mathbf{u}$). Vector $\mathbf{U}^{\text{obs}}$ will be called the observation vector. The $\mathbb{R}^{m_{\text{nobs}}}$-valued random vector $\mathbf{U}^{\text{nobs}} = (U_1^{\text{nobs}}, \ldots, U_{m_{\text{nobs}}}^{\text{nobs}})$ is made up of the m_{nobs} degrees of freedom (of the finite element model) for which no experimental data are available. A subset of these degrees of freedom will be used to perform the quality assessment of the identification. The random vector $\mathbf{U}$ appears as the unique deterministic nonlinear transformation of the finite family of N_p dependent random matrices $\{[\mathbb{A}(\mathbf{x})], \mathbf{x} \in \mathcal{I}\}$. This set of random matrices can then be represented by a $\mathbb{R}^{m_{\mathbb{V}}}$-valued random vector $\mathbb{V} = (V_1, \ldots, V_{m_\mathbb{V}})$. Consequently, the $\mathbb{R}^m$-valued random vector $\mathbf{U}$ can be written as

$$\mathbf{U} = \mathbf{h}(\mathbb{V}), \quad \mathbf{U}^{\text{obs}} = \mathbf{h}^{\text{obs}}(\mathbb{V}), \quad \mathbf{U}^{\text{nobs}} = \mathbf{h}^{\text{nobs}}(\mathbb{V}), \tag{8.2}$$

in which $v \mapsto \mathbf{h}(v) = (\mathbf{h}^{\text{obs}}(v), \mathbf{h}^{\text{nobs}}(v))$ is a deterministic nonlinear transformation from $\mathbb{R}^{m_\mathbb{V}}$ into $\mathbb{R}^m = \mathbb{R}^{m_{\text{obs}}} \times \mathbb{R}^{m_{\text{nobs}}}$ which can be constructed solving the discretized boundary value problem.

(3) Experimental data sets

It is assumed that ν_{exp} experimental data sets are available for the observation vector $\mathbf{U}^{\text{obs}}$. Each experimental data set corresponds to partial experimental data (only the trace of the displacement field on Γ_{obs} is observed) with a limited length (ν_{exp} is relatively small). These ν_{exp} experimental data sets correspond to measurements of ν_{exp} experimental configurations associated with the same boundary value problem. For configuration ℓ, with $\ell = 1, \ldots, \nu_{\text{exp}}$, the observation vector (corresponding to $\mathbf{U}^{\text{obs}}$ for the computational model) is denoted by $\mathbf{u}^{\text{exp},\ell}$ and belongs to $\mathbb{R}^m$. Therefore, the available data are made up of the ν_{exp} vectors $\mathbf{u}^{\text{exp},1}, \ldots, \mathbf{u}^{\text{exp},\nu_{\text{exp}}}$ in $\mathbb{R}^m$. Below, it is assumed that $\mathbf{u}^{\text{exp},1}, \ldots, \mathbf{u}^{\text{exp},\nu_{\text{exp}}}$ can be viewed as ν_{exp} independent realizations of a random vector $\mathbf{U}^{\text{exp}}$ defined on a probability space $(\Theta^{\text{exp}}, \mathcal{T}^{\text{exp}}, \mathcal{P}^{\text{exp}})$ and corresponding to random observation vector $\mathbf{U}^{\text{obs}}$ (but noting that random vectors $\mathbf{U}^{\text{exp}}$ and $\mathbf{U}^{\text{obs}}$ are not defined on the same probability space).

(4) Stochastic inverse problem to be solved

The problem to be solved concerns the identification of the unknown non-Gaussian random vector $\mathbb{V}$ representing the spatial discretization of fourth-order tensor-valued random field $\{\mathbb{C}(\mathbf{x}), \mathbf{x} \in \Omega\}$. Such an identification is carried out using partial and limited experimental data $\mathbf{u}^{\text{exp},1}$, $\ldots$, $\mathbf{u}^{\text{exp},\nu_{\text{exp}}}$ relative to the random observation vector $\mathbf{U}^{\text{obs}}$ such that $\mathbf{U}^{\text{obs}} = \mathbf{h}^{\text{obs}}(\mathbb{V})$ in which $\mathbf{h}^{\text{obs}}$ is a given deterministic nonlinear mapping. The components of the random vector $\mathbf{U}^{\text{nobs}}$, such that $\mathbf{U}^{\text{nobs}} = \mathbf{h}^{\text{nobs}}(\mathbb{V})$ in which $\mathbf{h}^{\text{nobs}}$ is a given deterministic nonlinear mapping, are not used for the identification but will be used for performing the quality assessment of the identification .

8.2 Construction of a family of prior algebraic probability models (PAPM) for the tensor-valued random field in elasticity theory

The notations introduced in Section 8.1 are used. We are interested in constructing a family of prior algebraic probability models (PAPM) for the non-Gaussian fourth-order tensor-valued random field $\{\mathbb{C}(\mathbf{x}), \mathbf{x} \in \Omega\}$ defined on a probability space $(\Theta, \mathcal{T}, \mathcal{P})$, in which $\mathbb{C}(\mathbf{x}) = \{\mathbb{C}_{ijk\ell}(\mathbf{x})\}_{ijk\ell}$.

8.2.1 Construction of the tensor-valued random field $\mathbb{C}$

The mean value $\{\underline{\mathbb{C}}(\mathbf{x}), \mathbf{x} \in \Omega\}$ of random field $\{\mathbb{C}(\mathbf{x}), \mathbf{x} \in \Omega\}$ is assumed to be given (this mean value can also be considered as unknown in the context of the inverse problem for the identification of the model parameters as proposed in Step 2 of Section 8.3). It is a deterministic tensor-valued field $\mathbf{x} \mapsto \{\underline{\mathbb{C}}_{ijkh}(\mathbf{x})\}_{ijkh}$ such that

$$E\{\mathbb{C}_{ijkh}(\mathbf{x})\} = \underline{\mathbb{C}}_{ijkh}(\mathbf{x}), \quad \forall \mathbf{x} \in \Omega, \tag{8.3}$$

where E is the mathematical expectation. For instance, in the context of the linear elasticity of a heterogeneous microstructure, tensor-valued function $\underline{\mathbb{C}}$ would be chosen as the mean model of a random anisotropic elastic microstructure at the mesoscale. It should be noted that the known symmetries, such as orthotropic symmetry or transversally isotropic symmetry, can be taken into account with the mean model represented by tensor $\{\underline{\mathbb{C}}_{ijkh}(\mathbf{x})\}_{ijkh}$. Nevertheless, in Section 8.2, the random fluctuations $\{\mathbb{C}_{ijkh}(\mathbf{x}) - \underline{\mathbb{C}}_{ijkh}(\mathbf{x})\}_{ijkh}$ around the mean tensor-valued field will be assumed to be purely anisotropic, without any symmetries. For instance, probability models for the elasticity tensor-valued random field with uncertain material symmetries are analyzed in (Guilleminot and Soize 2010). Below, we present an extension of the theory developed in (Soize 2006; 2008b). The family of prior tensor-valued random field $\mathbb{C}(\mathbf{x})$ is then constructed as

$$\mathbb{C}(\mathbf{x}) = C^0(\mathbf{x}) + \mathbf{C}(\mathbf{x}).$$

The deterministic tensor-valued field $C^0(\mathbf{x}) = \{C^0_{ijkh}(\mathbf{x})\}_{ijkh}$ will be called the "deterministic lower-bound tensor-valued field" which will be symmetric and positive definite. The tensor-valued random field $\mathbf{C}(\mathbf{x}) = \mathbf{C}_{ijkh}(\mathbf{x})\}_{ijkh}$ defined by $\mathbf{C}(\mathbf{x}) = \mathbb{C}(\mathbf{x}) - C^0(\mathbf{x})$ will be called the "fluctuations tensor-valued random field". This tensor will be almost surely symmetric and positive-definite. Tensor-valued field $C^0(\mathbf{x})$ should be such that the mean value $\underline{\mathbf{C}}(\mathbf{x}) = E\{\mathbf{C}(\mathbf{x})\} = \underline{\mathbb{C}}(\mathbf{x}) - C^0(\mathbf{x})$ is symmetric and positive definite. Below, we present the entire construction and we give the corresponding main mathematical properties.

(i)- Mathematical notations. In order to study the mathematical properties of the tensor-valued random field $\mathbb{C}(\mathbf{x})$, we introduce the real Hilbert space $\mathcal{H} = \{\mathbf{u} = (u_1, u_2, u_3)\,,\, u_j \in L^2(\Omega)\}$ equipped with the inner product

$$<\mathbf{u}\,,\mathbf{v}>_{\mathcal{H}}= \int_\Omega <\mathbf{u}(\mathbf{x})\,,\mathbf{v}(\mathbf{x})>\,d\mathbf{x},$$

and with the associated norm $\|\mathbf{u}\|_{\mathcal{H}} =<\mathbf{u}\,,\mathbf{u}>_{\mathcal{H}}^{1/2}$, in which $<\mathbf{u}(\mathbf{x})\,,\mathbf{v}(\mathbf{x})> = u_1(\mathbf{x})v_1(\mathbf{x}) + u_2(\mathbf{x})v_2(\mathbf{x}) + u_3(\mathbf{x})v_3(\mathbf{x})$ and where $L^2(\Omega)$ denotes the set of all the square integrable functions from Ω into $\mathbb{R}$. Let $\mathcal{V} \subset \mathcal{H}$ be the real Hilbert space representing the set of admissible displacement fields with values in $\mathbb{R}^3$ such that

$$\mathcal{V} = \{\mathbf{u} \in \mathcal{H}\ ,\ \partial\mathbf{u}/\partial x_1\,,\ \partial\mathbf{u}/\partial x_2\,,\ \partial\mathbf{u}/\partial x_3 \text{ in } \mathcal{H}\ ,\ \mathbf{u} = 0 \text{ on } \Gamma_0\},$$

equipped with the inner product

$$<\mathbf{u}\,,\mathbf{v}>_{\mathcal{V}} = <\mathbf{u}\,,\mathbf{v}>_{\mathcal{H}} + <\frac{\partial \mathbf{u}}{\partial x_j}\,,\frac{\partial \mathbf{v}}{\partial x_j}>_{\mathcal{H}}$$

and with the associated norm $\|\mathbf{u}\|_{\mathcal{V}} =<\mathbf{u}\,,\mathbf{u}>_{\mathcal{V}}^{1/2}$. The convention of summation over repeated latin indices is used. Let $L^2(\Theta, \mathcal{V})$ be the space of all the second-order random variable $\mathbf{U} = \{\mathbf{U}(\mathbf{x}), \mathbf{x} \in \Omega\}$ defined on $(\Theta, \mathcal{T}, \mathcal{P})$ with values in $\mathcal{V}$, equipped with the inner product

$$\ll \mathbf{U}, \mathbf{V} \gg = E\{<\mathbf{U}\,,\mathbf{V}>_{\mathcal{V}}\}$$

and the associated norm $|||\mathbf{U}||| =\ll \mathbf{U}, \mathbf{U} \gg^{1/2}$. For all $\mathbf{U}$ in $L^2(\Theta, \mathcal{V})$, we then have

$$|||\mathbf{U}|||^2 = E\{\|\mathbf{U}\|_{\mathcal{V}}^2\} < +\infty.$$

Finally, the operator norm $\|\mathbb{T}\|$ of any fourth-order tensor $\mathbb{T} = \{\mathbb{T}_{ijkh}\}_{ijkh}$ is defined by $\|\mathbb{T}\| = \sup_{\|z\|_F \leq 1} \|\mathbb{T}:z\|_F$ in which $z = \{z_{kh}\}_{kh}$ is a second-order tensor such that $\|z\|_F^2 = z_{kh}\, z_{kh}$ and where $\{\mathbb{T}:z\}_{ij} = \mathbb{T}_{ijkh} z_{kh}$.

(ii)- Mean tensor-valued field $\underline{\mathbb{C}}(\mathbf{x})$. We introduce the deterministic bilinear form $\underline{c}(\mathbf{u}, \mathbf{v})$ related to the mean tensor-valued field $\underline{\mathbb{C}}$,

$$\underline{c}(\mathbf{u}, \mathbf{v}) = \int_\Omega \underline{\mathbb{C}}_{ijkh}(\mathbf{x})\, \varepsilon_{kh}(\mathbf{u})\ \varepsilon_{ij}(\mathbf{v})\, d\mathbf{x}, \tag{8.4}$$

in which the second-order strain tensor $\{\varepsilon_{kh}\}_{kh}$ is such that

$$\varepsilon_{kh}(\mathbf{u}) = \frac{1}{2}\left(\frac{\partial u_k}{\partial x_h} + \frac{\partial u_h}{\partial x_k}\right).$$

For all $\mathbf{x}$, the fourth-order real tensor $\underline{\mathbb{C}}(\mathbf{x}) = \{\underline{\mathbb{C}}_{ijhk}(\mathbf{x})\}_{ijhk}$ of the elastic coefficients verifies the usual property of symmetry

$$\underline{\mathbb{C}}_{ijkh}(\mathbf{x}) = \underline{\mathbb{C}}_{jikh}(\mathbf{x}) = \underline{\mathbb{C}}_{ijhk}(\mathbf{x}) = \underline{\mathbb{C}}_{khij}(\mathbf{x}), \tag{8.5}$$

and for all symmetric second-order real tensor $z = \{z_{kh}\}_{kh}$, tensor $\underline{\mathbb{C}}(\mathbf{x})$ verifies the following property,

$$b_0\, z_{kh}\, z_{kh} \;\leq\; \underline{\mathbb{C}}_{ijkh}(\mathbf{x}) z_{kh} z_{ij} \;\leq\; b_1\, z_{kh}\, z_{kh}, \tag{8.6}$$

in which b_0 and b_1 are deterministic positive constants which are independent of $\mathbf{x}$. Taking into account Eqs. (8.5) and (8.6), it can be deduced that bilinear form $\underline{c}(\mathbf{u}, \mathbf{v})$ is symmetric, positive-definite, continuous on $\mathcal{V} \times \mathcal{V}$ and is elliptic on $\mathcal{V}$, that is to say, is such that

$$\underline{c}(\mathbf{u}, \mathbf{u}) \geq k_0\, \|\mathbf{u}\|^2_{\mathcal{V}}. \tag{8.7}$$

Equation (8.7) can easily be proven using Eq. (8.4), Eq. (8.6) and the Korn inequality which is written as $\int_\Omega \varepsilon_{kh}(\mathbf{u})\, \varepsilon_{kh}(\mathbf{u})\, d\mathbf{x} \geq b_2\, \|\mathbf{u}\|^2_{\mathcal{V}}$. It can then be deduced that $k_0 = b_0\, b_2$ is a positive constant.

(iii)- Deterministic lower-bound tensor-valued field $C^0(\mathbf{x})$. The deterministic lower-bound tensor-valued field $\{C^0(\mathbf{x}), \mathbf{x} \in \Omega\}$ is a given deterministic field which is introduced to guaranty the ellipticity condition of the tensor-valued random field $\mathbb{C}(\mathbf{x})$. We will give two examples for the construction of $C^0(\mathbf{x})$. The fourth-order real tensor $C^0_{ijkh}(\mathbf{x})$ must verify the usual property of symmetry (similarly to Eq. (8.5)) and for all symmetric second-order real tensor $\{z_{kh}\}_{kh}$ must be such that

$$b^0_0\, z_{kh}\, z_{kh} \;\leq\; C^0_{ijkh}(\mathbf{x}) z_{kh}\, z_{ij} \;\leq\; b^0_1\, z_{kh}\, z_{kh}, \quad \forall \mathbf{x} \in \Omega, \tag{8.8}$$

in which b^0_0 and b^0_1 are deterministic positive constants which are independent of $\mathbf{x}$. Let $\underline{\mathbf{C}}(\mathbf{x})$ be the tensor-valued deterministic field defined by

$$\underline{\mathbf{C}}(\mathbf{x}) = \underline{\mathbb{C}}(\mathbf{x}) - C^0(\mathbf{x}), \quad \forall \mathbf{x} \in \Omega. \tag{8.9}$$

In addition, tensor-valued field C^0 must be constructed for that, for all $\mathbf{x}$, tensor $\underline{\mathbf{C}}(\mathbf{x})$, which verifies the symmetry property (see Eq. (8.5)), must be positive definite, that is to say, for all non zero symmetric second-order real tensor $\{z_{kh}\}_{kh}$, must be such that

$$\underline{\mathbf{C}}_{ijkh}(\mathbf{x})\, z_{kh}\, z_{ij} > 0, \quad \forall \mathbf{x} \in \Omega. \tag{8.10}$$

From Eqs. (8.9), (8.6) and (8.8), it can easily be deduced that

$$\underline{\mathbf{C}}_{ijkh}(\mathbf{x})\, z_{kh}\, z_{ij} \;\leq\; b_1^1\, z_{kh}\, z_{kh}, \quad \forall \mathbf{x} \in \Omega, \tag{8.11}$$

in which $b_1^1 = b_1 - b_0^0 > 0$ is a positive finite constant independent of $\mathbf{x}$. Introducing the deterministic bilinear form $\underline{c}^0(\mathbf{u}, \mathbf{v})$ related to the deterministic lower-bound tensor-valued field $C^0(\mathbf{x})$,

$$\underline{c}^0(\mathbf{u}, \mathbf{v}) = \int_\Omega C^0_{ijkh}(\mathbf{x})\, \varepsilon_{kh}(\mathbf{u})\; \varepsilon_{ij}(\mathbf{v})\, d\mathbf{x}, \tag{8.12}$$

it can be shown, as previously, that this bilinear form is symmetric, positive-definite, continuous on $\mathcal{V} \times \mathcal{V}$ and is elliptic on $\mathcal{V}$, that is to say, is such that

$$\underline{c}^0(\mathbf{u}, \mathbf{u}) \geq k_0^0\, \|\mathbf{u}\|^2_{\mathcal{V}}, \tag{8.13}$$

in which $k_0^0 = b_0^0\, b_2$ is a positive constant.

Example 1. In certain cases, a deterministic lower bound $C^{\min}$ independent of $\mathbf{x}$ can be constructed for a given microstructure (Guilleminot et al. 2011). The fourth-order tensor $C^{\min}$ is symmetric and positive definite. For all $\mathbf{x}$ in Ω, we then have $C^0(\mathbf{x}) = C^{\min}$.

Example 2. If there is no available information to construct the deterministic lower-bound tensor-valued field $\{C^0(\mathbf{x}), \mathbf{x} \in \Omega\}$, we can define it as $C^0(\mathbf{x}) = \epsilon_0\, \underline{\mathbf{C}}(\mathbf{x})$ in which $0 < \epsilon_0 < 1$ can be chosen as small as one wants. With such a choice, we have $\underline{\mathbf{C}}(\mathbf{x}) = (1 - \epsilon_0)\underline{\mathbb{C}}(\mathbf{x})$.

(iv)- Random fluctuations tensor-valued field $\mathbf{C}(\mathbf{x})$. The random fluctuations tensor-valued field $\{\mathbf{C}(\mathbf{x}), \mathbf{x} \in \Omega\}$ is defined on probability space $(\Theta, \mathcal{T}, \mathcal{P})$. In (Soize 2006; 2008b), the random fluctuations tensor-valued field $\{\mathbf{C}(\mathbf{x}), \mathbf{x} \in \Omega\}$ is constructed in order that all the following properties listed below be verified.
For all $\mathbf{x}$ fixed in Ω, the fourth-order real tensor $\mathbf{C}_{ijkh}(\mathbf{x})$ is symmetric,

$$\mathbf{C}_{ijkh}(\mathbf{x}) = \mathbf{C}_{jikh}(\mathbf{x}) = \mathbf{C}_{ijhk}(\mathbf{x}) = \mathbf{C}_{khij}(\mathbf{x}), \tag{8.14}$$

and is positive definite, that is to say, for all non zero symmetric second-order real tensor $\{z_{kh}\}_{kh}$, we have,

$$\mathbf{C}_{ijkh}(\mathbf{x})\, z_{kh}\, z_{ij} > 0. \tag{8.15}$$

The mean function of random field $\{\mathbf{C}(\mathbf{x}), \mathbf{x} \in \Omega\}$ is equal to the tensor-valued deterministic field $\{\underline{\mathbf{C}}(\mathbf{x}), \mathbf{x} \in \Omega\}$ defined by Eq. (8.9),

$$E\{\mathbf{C}(\mathbf{x})\} = \underline{\mathbf{C}}(\mathbf{x}), \quad \forall \mathbf{x} \in \Omega. \tag{8.16}$$

Let $\mathbf{c}(\mathbf{U},\mathbf{V})$ be the random bilinear form defined by

$$\mathbf{c}(\mathbf{U},\mathbf{V}) = \int_\Omega \mathbf{C}_{ijkh}(\mathbf{x})\,\varepsilon_{kh}(\mathbf{U})\;\varepsilon_{ij}(\mathbf{V})\,d\mathbf{x}. \tag{8.17}$$

The available information used to construct the random field $\mathbf{C}$ implies (see (Soize 2006)) that the bilinear form $(\mathbf{U},\mathbf{V}) \mapsto E\{\mathbf{c}(\mathbf{U},\mathbf{V})\}$ is symmetric, positive definite, continuous on $L^2(\Theta,\mathcal{V}) \times L^2(\Theta,\mathcal{V})$, is not elliptic but is such that, for all $\mathbf{U}$ in $L^2(\Theta,\mathcal{V})$, we have

$$\sqrt{E\{\mathbf{c}(\mathbf{U},\mathbf{U})^2\}} \geq k_1\, E\{\|\mathbf{U}\|^2_{\mathcal{V}}\}, \tag{8.18}$$

in which k_1 is a positive constant. Equation (8.18) implies that the following elliptic boundary value problem $E\{\mathbf{c}(\mathbf{U},\mathbf{V})\} = E\{f(\mathbf{V})\}$ for all $\mathbf{V}$ in $L^2(\Theta,\mathcal{V})$, in which $f(\mathbf{v})$ is a given continuous linear form on $\mathcal{V}$, has a unique random solution $\mathbf{U}$ in $L^2(\Theta,\mathcal{V})$, but the random solution $\mathbf{U}$ is not a continuous function of the parameters.

(v)- Prior algebraic probability model (PAPM) for the tensor-valued random field $\mathbb{C}(\mathbf{x})$. The non-Gaussian fourth-order tensor-valued random field $\{\mathbb{C}(\mathbf{x}), \mathbf{x} \in \Omega\}$ is defined on probability space $(\Theta, \mathcal{T}, \mathcal{P})$ and such that, for all $\mathbf{x}$ in Ω,

$$\mathbb{C}(\mathbf{x}) = C^0(\mathbf{x}) + \mathbf{C}(\mathbf{x}), \quad \forall \mathbf{x} \in \Omega, \tag{8.19}$$

in which the deterministic lower-bound tensor-valued field $\{C^0(\mathbf{x}), \mathbf{x} \in \Omega\}$ is defined in (iii) and where the random fluctuations tensor-valued field $\{\mathbf{C}(\mathbf{x}), \mathbf{x} \in \Omega\}$ is defined in (iv). Let $\mathbb{c}(\mathbf{U},\mathbf{V})$ be the random bilinear form defined by

$$\mathbb{c}(\mathbf{U},\mathbf{V}) = \int_\Omega \mathbb{C}_{ijkh}(\mathbf{x})\,\varepsilon_{kh}(\mathbf{U})\;\varepsilon_{ij}(\mathbf{V})\,d\mathbf{x}, \tag{8.20}$$

and let $c(\mathbf{U},\mathbf{V})$ be the bilinear form defined by

$$c(\mathbf{U},\mathbf{V}) = E\{\mathbb{c}(\mathbf{U},\mathbf{V})\}. \tag{8.21}$$

Then, it can easily be verified that the bilinear form $c(\mathbf{U},\mathbf{V})$ is symmetric, positive-definite, continuous on $L^2(\Theta,\mathcal{V}) \times L^2(\Theta,\mathcal{V})$ and is elliptic, that is to say, for all $\mathbf{U}$ in $L^2(\Theta,\mathcal{V})$, we have

$$c(\mathbf{U},\mathbf{U}) \geq k_0^0\, |||\mathbf{U}|||^2. \tag{8.22}$$

Equation (8.22) implies that the following elliptic boundary value problem $E\{\mathbb{c}(\mathbf{U},\mathbf{V})\} = E\{f(\mathbf{V})\}$ for all $\mathbf{V}$ in $L^2(\Theta,\mathcal{V})$, in which $f(\mathbf{v})$ is a given continuous linear form on $\mathcal{V}$, has a unique random solution $\mathbf{U}$ in $L^2(\Theta,\mathcal{V})$ and the random solution $\mathbf{U}$ is a continuous function of the parameters.

8.2.2 Construction of the tensor-valued random field $\mathbf{C}$

In this section, we summarize the construction of random field $\{\mathbf{C}(\mathbf{x}), \mathbf{x} \in \Omega\}$ whose available information and properties have been defined in Section 8.2.1-(iv) and for which the details of this construction can be found in (Soize 2008b) and (Soize 2006). For all $\mathbf{x}$ fixed in $\mathcal{I} = \{\mathbf{x}^1, \ldots, \mathbf{x}^{N_p}\} \subset \Omega$, the fourth-order tensor-valued random variable $\mathbf{C}(\mathbf{x})$ is represented by a real random matrix $[\mathbf{A}(\mathbf{x})]$. Let I and J be the new indices belonging to $\{1, \ldots, 6\}$ such that $I = (i,j)$ and $J = (k,h)$ with the following correspondence: $1 = (1,1), 2 = (2,2), 3 = (3,3), 4 = (1,2), 5 = (1,3)$ and $6 = (2,3)$. Thus, for all $\mathbf{x}$ in Ω, the random (6×6) real matrix $[\mathbf{A}(\mathbf{x})]$ is such that

$$[\mathbf{A}(\mathbf{x})]_{IJ} = \mathbf{C}_{ijkh}(\mathbf{x}), \quad \mathbf{x} \in \Omega. \tag{8.23}$$

For all $\mathbf{x}$ fixed in Ω, due to the symmetry and positive-definiteness properties (defined by Eqs. (8.14) and (8.15)) of the random fourth-order tensor $\mathbf{C}(\mathbf{x})$, it can be deduced that $[\mathbf{A}(\mathbf{x})]$ is a random variable with values in the set $\mathbb{M}_6^+(\mathbb{R})$ of all the (6×6) real symmetric positive-definite matrices. The $\mathbb{M}_6^+(\mathbb{R})$-valued random field $\{[\mathbf{A}(\mathbf{x})], \mathbf{x} \in \Omega\}$, indexed by Ω, defined on the probability space $(\Theta, \mathcal{T}, \mathcal{P})$, is constituted of $6 \times (6+1)/2 = 21$ mutually dependent real-valued random fields defining the fourth-order tensor-valued random field $\mathbf{C}$ indexed by Ω. Let $\mathbf{x} \mapsto [\underline{a}(\mathbf{x})]$ be the random field from Ω into $\mathbb{M}_6^+(\mathbb{R})$ defined by

$$[\underline{a}(\mathbf{x})]_{IJ} = \underline{\mathbf{C}}_{ijkh}(\mathbf{x}), \quad \mathbf{x} \in \Omega, \tag{8.24}$$

in which the random field $\mathbf{x} \mapsto \underline{\mathbf{C}}(\mathbf{x})$ is defined by Eq. (8.9). Consequently, the mean function of random field $[\mathbf{A}]$ is such that

$$E\{[\mathbf{A}(\mathbf{x})]\} = [\underline{a}(\mathbf{x})], \quad \mathbf{x} \in \Omega. \tag{8.25}$$

Since $[\underline{a}(\mathbf{x})]$ belongs to $\mathbb{M}_6^+(\mathbb{R})$, there is an upper triangular invertible matrix $[\underline{L}(\mathbf{x})]$ in $\mathbb{M}_6(\mathbb{R})$ (the set of all the (6×6) real matrices) such that

$$[\underline{a}(\mathbf{x})] = [\underline{L}(\mathbf{x})]^T \, [\underline{L}(\mathbf{x})], \quad \mathbf{x} \in \Omega. \tag{8.26}$$

From Eq. (8.11), $\mathbf{x} \mapsto [\underline{a}(\mathbf{x})]$ is a bounded function on Ω and it can then be assumed that $\mathbf{x} \mapsto [\underline{L}(\mathbf{x})]$ is also bounded function on Ω. For all $\mathbf{x}$ fixed in Ω, the random matrix $[\mathbf{A}(\mathbf{x})]$ can be written as

$$[\mathbf{A}(\mathbf{x})] = [\underline{L}(\mathbf{x})]^T \, [\mathbf{G}_0(\mathbf{x})] \, [\underline{L}(\mathbf{x})], \tag{8.27}$$

in which $\mathbf{x} \mapsto [\mathbf{G}_0(\mathbf{x})]$ is a random field defined on $(\Theta, \mathcal{T}, \mathcal{P})$, indexed by $\mathbb{R}^3$, with values in $\mathbb{M}_6^+(\mathbb{R})$, such that for all $\mathbf{x}$ in $\mathbb{R}^3$

$$E\{[\mathbf{G}_0(\mathbf{x})]\} = [\,I_6], \tag{8.28}$$

in which $[I_6]$ is the unity matrix. The random field $[\mathbf{G}_0]$ is completely defined below.

(i)- Probability model of the random field $[\mathbf{G}_0]$. The random field $\mathbf{x} \mapsto [\mathbf{G}_0(\mathbf{x})]$ is constructed as a homogeneous and normalized non-Gaussian positive-definite matrix-valued random field, defined on probability space $(\Theta, \mathcal{T}, \mathcal{P})$, indexed by $\mathbb{R}^3$, with values in $\mathbb{M}_6^+(\mathbb{R})$. This random field is constructed as a non-linear mapping of 21 independent second-order centered homogeneous Gaussian random fields $\mathbf{x} \mapsto U_{jj'}(\mathbf{x})$, $1 \leq j \leq j' \leq 6$, defined on probability space $(\Theta, \mathcal{T}, \mathcal{P})$, indexed by $\mathbb{R}^3$, with values in $\mathbb{R}$, and named the stochastic germs of the non-Gaussian random field $[\mathbf{G}_0]$.

(i.1)- Random fields $U_{jj'}$ as the stochastic germs of the random field $[\mathbf{G}_0]$. The stochastic germs are constituted of 21 independent second-order centered homogeneous Gaussian random fields $\mathbf{x} \mapsto U_{jj'}(\mathbf{x})$, $1 \leq j \leq j' \leq 6$, defined on probability space $(\Theta, \mathcal{T}, \mathcal{P})$, indexed by $\mathbb{R}^3$, with values in $\mathbb{R}$ and such that

$$E\{U_{jj'}(\mathbf{x})\} = 0, \quad E\{U_{jj'}(\mathbf{x})^2\} = 1. \tag{8.29}$$

Consequently, all these random fields are completely and uniquely defined by the 21 autocorrelation functions $R_{U_{jj'}}(\boldsymbol{\zeta}) = E\{U_{jj'}(\mathbf{x} + \boldsymbol{\zeta})\, U_{jj'}(\mathbf{x})\}$ defined for all $\boldsymbol{\zeta} = (\zeta_1, \zeta_2, \zeta_3)$ in $\mathbb{R}^3$ and such that $R_{U_{jj'}}(0) = 1$. In order to obtain a class having a reasonable number of parameters, these autocorrelation functions are written as $R_{U_{jj'}}(\boldsymbol{\zeta}) = \rho_1^{jj'}(\zeta_1) \times \rho_2^{jj'}(\zeta_2) \times \rho_3^{jj'}(\zeta_3)$ in which, for all $k = 1, 2, 3$, one has $\rho_k^{jj'}(0) = 1$ and for all $\eta_k \neq 0$,

$$E\rho_k^{jj'}(\zeta_k) = 4(L_k^{jj'})^2/(\pi^2\zeta_k^2)\ \sin^2\!\Big(\pi\zeta_k/(2L_k^{jj'})\Big)\,. \tag{8.30}$$

in which $L_1^{jj'}, L_2^{jj'}, L_3^{jj'}$ are positive real numbers. Each random field $U_{jj'}$ is then mean-square continuous on $\mathbb{R}^3$ and it power spectral measure has a compact support. Such a model has 63 real parameters $L_1^{jj'}, L_2^{jj'}, L_3^{jj'}$ for $1 \leq j \leq j' \leq 6$ which represent the spatial correlation lengths of the stochastic germs $U_{jj'}$.

(i.2)- Defining an adapted family of functions. The construction of the random field $[\mathbf{G}_0]$ requires the introduction of an adapted family of functions $\{u \mapsto h(\alpha, u)\}_{\alpha > 0}$ in which α is a positive real number. Function $u \mapsto h(\alpha, u)$, from $\mathbb{R}$ into $]0\,, +\infty[$, is introduced such that $\Gamma_\alpha = h(\alpha, U)$ is a gamma random variable with parameter α while U is a normalized

Gaussian random variable ($E\{U\} = 0$ and $E\{U^2\} = 1$). Consequently, for all u in $\mathbb{R}$, one has

$$h(\alpha, u) = F_{\Gamma_\alpha}^{-1}(F_U(u)), \tag{8.31}$$

in which $u \mapsto F_U(u) = \int_{-\infty}^{u} \frac{1}{\sqrt{2\pi}} e^{-v}\, dv$ is the cumulative distribution function of the normalized Gaussian random variable U. The function $p \mapsto F_{\Gamma_\alpha}^{-1}(p)$ from $]0\,,1[$ into $]0\,,+\infty[$ is the reciprocal function of the cumulative distribution function $\gamma \mapsto F_{\Gamma_\alpha}(\gamma) = \int_0^{\gamma} \frac{1}{\Gamma(\alpha)}\, t^{\alpha-1}\, e^{-t}\, dt$ of the gamma random variable Γ_α with parameter α in which $\Gamma(\alpha)$ is the gamma function defined by $\Gamma(\alpha) = \int_0^{+\infty} t^{\alpha-1}\, e^{-t}\, dt$.

(i.3)- Defining the random field $[\mathbf{G}_0]$. For all $\mathbf{x}$ fixed in Ω, the available information defined by Eqs. (8.23) to (8.28), lead us to choose the random matrix $[\mathbf{G}_0(\mathbf{x})]$ in ensemble SG_0^+ defined in Section 2.5.2. Taking into account the properties defined in Section 2.5-(1), (2) and (3), the correlation spatial structure of random field $\mathbf{x} \mapsto [\mathbf{G}_0(\mathbf{x})]$ is then introduced in replacing the Gaussian random variables $U_{jj'}$ by the Gaussian real-valued random fields $\{U_{jj'}(\mathbf{x}), \mathbf{x} \in \mathbb{R}^3\}$ defined in Section 8.2.2-(i.1), for which the correlation spatial structure is defined by a spatial correlation lengths $L_k^{jj'}$. Consequently, the random field $\mathbf{x} \mapsto [\mathbf{G}_0(\mathbf{x})]$, defined on probability space $(\Theta, \mathcal{T}, \mathcal{P})$, indexed by $\mathbb{R}^3$, with values in $\mathbb{M}_6^+(\mathbb{R})$ is constructed as follows:
(1) Let $\{U_{jj'}(\mathbf{x}), \mathbf{x} \in \mathbb{R}^3\}_{1 \le j \le j' \le 6}$ be the 21 independent random fields introduced in Section 8.2.2-(i.1). Consequently, for all $\mathbf{x}$ in $\mathbb{R}^3$,

$$E\{U_{jj'}(\mathbf{x})\} = 0, \quad E\{U_{jj'}(\mathbf{x})^2\} = 1, \quad 1 \le j \le j' \le 6. \tag{8.32}$$

(2) Let δ be the real number, independent of $\mathbf{x}$, such that

$$0 < \delta < \sqrt{7/11} \; < \; 1. \tag{8.33}$$

This parameter which is assumed to be known (resulting, for instance, from an experimental identification solving an inverse problem) allows the statistical fluctuations (dispersion) of the random field $[\mathbf{G}_0]$ to be controlled.
(3) For all $\mathbf{x}$ in $\mathbb{R}^3$, the random matrix $[\mathbf{G}_0(\mathbf{x})]$ is written as

$$[\mathbf{G}_0(\mathbf{x})] = [\mathbf{L}(\mathbf{x})]^T\, [\mathbf{L}(\mathbf{x})], \tag{8.34}$$

in which $[\mathbf{L}(\mathbf{x})]$ is the upper (6×6) real triangular random matrix defined (see Section 2.5.2) as follows:
$\triangleright$ For $1 \le j \le j' \le 6$, the 21 random fields $\mathbf{x} \mapsto [\mathbf{L}(\mathbf{x})]_{jj'}$ are independent.
$\triangleright$ For $j < j'$, the real-valued random field $\mathbf{x} \mapsto [\mathbf{L}(\mathbf{x})]_{jj'}$, indexed by $\mathbb{R}^3$,

is defined by $[\mathbf{L}(\mathbf{x})]_{jj'} = \sigma\, U_{jj'}(\mathbf{x})$ in which σ is such that $\sigma = \delta/\sqrt{7}$.
$\triangleright$ For $j = j'$, the positive-valued random field $\mathbf{x} \mapsto [\mathbf{L}(\mathbf{x})]_{jj}$, indexed by $\mathbb{R}^3$, is defined by $[\mathbf{L}(\mathbf{x})]_{jj} = \sigma\,\sqrt{2\,h(\alpha_j, U_{jj}(\mathbf{x}))}$ in which $\alpha_j = 7/(2\delta^2) + (1-j)/2$.

(i.4)- A few basic properties of the random field $[\mathbf{G}_0]$. The random field $\mathbf{x} \mapsto [\mathbf{G}_0(\mathbf{x})]$, defined in Section 8.2.2-(i.3), is a homogeneous second-order mean-square continuous random field indexed by $\mathbb{R}^3$ with values in $\mathbb{M}_6^+(\mathbb{R})$ and its trajectories are almost surely continuous on $\mathbb{R}^3$. For all $\mathbf{x} \in \mathbb{R}^3$, one has

$$E\{\|\mathbf{G}_0(\mathbf{x})\|_F^2\} < +\infty, \quad E\{[\mathbf{G}_0(\mathbf{x})]\} = [\,I_6]. \tag{8.35}$$

It can be proven that the newly introduced parameter δ corresponds to the following definition

$$\delta = \left\{\frac{1}{6}E\{\|\,[\mathbf{G}_0(\mathbf{x})] - [\,I_6]\,\|_F^2\}\right\}^{1/2}, \tag{8.36}$$

which shows that

$$E\{\|\,\mathbf{G}_0(\mathbf{x})\,\|_F^2\} = 6\,(\delta^2 + 1), \tag{8.37}$$

in which δ is independent of $\mathbf{x}$. For all $\mathbf{x}$ fixed in $\mathbb{R}^3$, the probability density function with respect to the measure $\widetilde{d}G = 2^{15/2}\,\Pi_{1\leq j\leq k\leq 6}\, d[G]_{jk}$ of random matrix $[\mathbf{G}_0(\mathbf{x})]$ is independent of $\mathbf{x}$ and is written (see Section 2.5.2 with $n = 6$) as

$$p_{[\mathbf{G}_0(\mathbf{x})]}([G]) = \mathbb{1}_{\mathbb{M}_6^+(\mathbb{R})}([G]) \times C_{\mathbf{G}_0} \times \left(\det [G]\right)^{7\frac{(1-\delta^2)}{2\delta^2}} \times \exp\left\{-\frac{7}{2\delta^2}\,\mathrm{tr}\,[G]\right\}, \tag{8.38}$$

where the positive constant $C_{\mathbf{G}_0}$ is defined in Section 2.5.2 with $n = 6$. For all $\mathbf{x}$ fixed in $\mathbb{R}^3$, Eq. (8.38) shows that the random variables $\{[\mathbf{G}_0(\mathbf{x})]_{jk},\ 1 \leq j \leq k \leq 6\}$ are mutually dependent. In addition, the system of the marginal probability distributions of the random field $\mathbf{x} \mapsto [\mathbf{G}_0(\mathbf{x})]$ is completely defined and is not Gaussian. There exists a positive constant b_G independent of $\mathbf{x}$, but depending on δ, such that for all $\mathbf{x} \in \mathbb{R}^3$,

$$E\{\|[\mathbf{G}_0(\mathbf{x})]^{-1}\|^2\} \leq b_G < +\infty. \tag{8.39}$$

Since $[\mathbf{G}_0(\mathbf{x})]$ is a random matrix with values in $\mathbb{M}_6^+(\mathbb{R})$, then $[\mathbf{G}_0(\mathbf{x})]^{-1}$ exists (almost surely). However, since almost sure convergence does not imply mean-square convergence, the previous result cannot simply be deduced. Let $\overline{\Omega} = \Omega \cup \partial\Omega$ be the closure of the bounded set Ω. We then have

$$E\{(\sup_{\mathbf{x}\in\overline{\Omega}}\|\,[\mathbf{G}_0(\mathbf{x})]^{-1}\|)^2\} = c_G^2 \;<\; +\infty. \tag{8.40}$$

in which sup is the supremum and where $0 < c_G < +\infty$ is a finite positive constant.

(ii)- Properties of the random field $[\mathbf{A}]$. The random field $\mathbf{x} \mapsto [\mathbf{A}(\mathbf{x})]$ indexed by Ω, with values in $\mathbb{M}_6^+(\mathbb{R})$, is defined by Eq. (8.27) in which the random field $\mathbf{x} \mapsto [\mathbf{G}_0(\mathbf{x})]$ indexed by $\mathbb{R}^3$ with values in $\mathbb{M}_6^+(\mathbb{R})$, is defined in Section 8.2.2-(i).

(ii.1)- Basic properties of the random field $[\boldsymbol{A}]$ *and its parameters.* The random field $\mathbf{x} \mapsto [\mathbf{A}(\mathbf{x})]$ is a second-order random field on Ω,

$$E\{\|\mathbf{A}(\mathbf{x})\|^2\} \leq E\{\|\mathbf{A}(\mathbf{x})\|_F^2\} < +\infty. \tag{8.41}$$

The system of the marginal probability distributions of the random field $\mathbf{x} \mapsto [\mathbf{A}(\mathbf{x})]$ is completely defined, is not Gaussian and is deduced from the system of the marginal probability distributions of the random field $\mathbf{x} \mapsto [\mathbf{G}_0(\mathbf{x})]$ by using Eq. (8.27). In general, since $[\underline{a}(\mathbf{x})]$ depends on $\mathbf{x}$, then the random field $\{[\mathbf{A}(\mathbf{x})]\,, \mathbf{x} \in \Omega\}$ is non homogeneous. It can easily be proven that

$$E\{\|[\mathbf{A}(\mathbf{x})] - [\underline{a}(\mathbf{x})]\|_F^2\} = \frac{\delta^2}{(n+1)}\{\|\underline{a}(\mathbf{x})\|_F^2 + (\operatorname{tr}[\underline{a}(\mathbf{x})])^2\}.$$

The dispersion parameter $\delta_A(\mathbf{x})$ is such that

$$\delta_A(\mathbf{x})^2 = \frac{E\{\|[\mathbf{A}(\mathbf{x})] - [\underline{a}(\mathbf{x})]\|_F^2\}}{\|\underline{a}(\mathbf{x})\|_F^2}, \tag{8.42}$$

and can be rewritten as

$$\delta_A(\mathbf{x}) = \frac{\delta}{\sqrt{7}}\left\{1 + \frac{(\operatorname{tr}[\underline{a}(\mathbf{x})])^2}{\operatorname{tr}\{[\underline{a}(\mathbf{x})]^2\}}\right\}^{1/2}. \tag{8.43}$$

Random field $\mathbf{x} \mapsto [\mathbf{G}_0(\mathbf{x})]$ has almost surely continuous trajectories (see Section 8.2.2-(i.4)). If function $\mathbf{x} \mapsto [\underline{a}(\mathbf{x})]$ is continuous on $\overline{\Omega}$, then random field $\mathbf{x} \mapsto [\mathbf{A}(\mathbf{x})]$ has almost surely continuous trajectories on $\overline{\Omega}$ (if not, it is not true). Equations (8.23), (8.24), (8.27) et (8.39), (8.40) allow Eq. (8.18) to be proven.

Random field $\mathbf{x} \mapsto [\mathbf{A}(\mathbf{x})]$ is completely and uniquely defined by the following parameters: the $\mathbb{M}_6^+(\mathbb{R})$-valued mean function $\mathbf{x} \mapsto [\underline{a}(\mathbf{x})]$, the positive real parameter δ and the 63 positive real parameters $L_1^{jj'}, L_2^{jj'}, L_3^{jj'}$ for $1 \leq j \leq j' \leq 6$. The smallest number of parameters corresponds to the following case: $\mathbf{x} \mapsto [\underline{a}(\mathbf{x})]$, δ and $L_d = L_1^{jj'} = L_2^{jj'} = L_3^{jj'}$ for all $1 \leq j \leq j' \leq 6$.

(ii.2)- Spatial correlation lengths of the random field $[\mathbf{A}]$ *for the homogeneous case.* If $[\underline{a}(\mathbf{x})] = [\underline{a}]$ is independent of $\mathbf{x}$, then the random field $\{[\mathbf{A}(\mathbf{x})] = [\underline{L}]^T [\mathbf{G}_0(\mathbf{x})] [\underline{L}], \mathbf{x} \in \Omega\}$ can be viewed as the restriction to Ω of a homogeneous random field indexed by $\mathbb{R}^3$. Then the dispersion parameter given by Eq. (8.43) is independent of $\mathbf{x}$ and then $\delta_A(\mathbf{x}) = \delta_A$. Let $\boldsymbol{\zeta} = (\zeta_1, \zeta_2, \zeta_3) \mapsto r^A(\boldsymbol{\zeta})$ be the function defined from $\mathbb{R}^3$ into $\mathbb{R}$ by

$$r^A(\boldsymbol{\zeta}) = \frac{\operatorname{tr} E\{([\mathbf{A}(\mathbf{x}+\boldsymbol{\zeta})] - [\underline{a}])\,([\mathbf{A}(\mathbf{x})] - [\underline{a}])\}}{E\{\|[\mathbf{A}(\mathbf{x})] - [\underline{a}]\|_F^2\}}. \tag{8.44}$$

It can be seen that $r^A(0) = 1$ and $r^A(-\boldsymbol{\zeta}) = r^A(\boldsymbol{\zeta})$. For $k = 1, 2, 3$, the spatial correlation length L_k^A of $\mathbf{x} \mapsto [\mathbf{A}(\mathbf{x})]$, relative to the coordinate x_k, can then be defined by

$$L_k^A = \int_0^{+\infty} |r^A(\boldsymbol{\zeta}^k)|\, d\zeta_k, \tag{8.45}$$

in which $\boldsymbol{\zeta}^1 = (\zeta_1, 0, 0)$, $\boldsymbol{\zeta}^2 = (0, \zeta_2, 0)$ and $\boldsymbol{\zeta}^3 = (0, 0, \zeta_3)$. It is also possible to define a fourth-order tensor $L_k^{A_{\alpha\beta\gamma\kappa}}$ of spatial correlation lengths relative to the coordinate x_k such that

$$L_k^{A_{\alpha\beta\gamma\kappa}} = \int_0^{+\infty} |r^{A_{\alpha\beta\gamma\kappa}}(\boldsymbol{\zeta}^k)|\, d\zeta_k, \quad k = 1, 2, 3, \tag{8.46}$$

in which

$$r^{A_{\alpha\beta\gamma\kappa}}(\boldsymbol{\zeta}) = \frac{1}{\sigma_{\alpha\beta}\sigma_{\gamma\kappa}} E\{([\mathbf{A}(\mathbf{x}+\boldsymbol{\zeta})]_{\alpha\beta} - [\underline{a}]_{\alpha\beta})\,([\mathbf{A}(\mathbf{x})]_{\gamma\kappa} - [\underline{a}]_{\gamma\kappa})\}, \tag{8.47}$$

and where $\sigma_{\alpha\beta} = \sqrt{E\{([\mathbf{A}(\mathbf{x})]_{\alpha\beta} - [\underline{a}]_{\alpha\beta})^2\}}$.

8.3 Methodology for the identification of a high-dimension polynomial chaos expansion using partial and limited experimental data

The identification of Bayesian posteriors of a high-dimension Polynomial Chaos Expansion (PCE) with random Vector-Valued Coefficients (VVC), using partial and limited experimental data, requires a first identification of the deterministic VVC of the high-dimension PCE for the non-Gaussian tensor-valued random field. Such a first identification, performed in four steps, is described in details in (Soize 2010b). Below, we briefly summarize this methodology.

Taking into account the problem defined in Section 8.1, since we have $\mathbb{C}_{ijkh}(\mathbf{x}) = [\mathbb{A}(\mathbf{x})]_{IJ}$ for all $\mathbf{x}$ in Ω, we have to identify the probability distribution of the $\mathbb{R}^{m_{\mathbb{V}}}$-valued random vector $\mathbb{V} = (V_1, \ldots, V_{m_{\mathbb{V}}})$ which represents the finite family of N_p dependent random matrices $\{[\mathbb{A}(\mathbf{x})], \mathbf{x} \in \mathcal{I}\}$.

Step 1. *Introduction of a family of Prior Algebraic Probability Models (PAPM) for random vector* $\mathbb{V}$. The available partial and limited experimental data are not sufficient to perform a direct statistical estimation of the covariance matrix $[C_{\mathbb{V}}]$, that would be necessary to construct a reduced-order statistical model deduced from the Karhunen-Loeve expansion of the random field $\{[\mathbb{A}(\mathbf{x})], \mathbf{x} \in \Omega\}$ (that is to say deduced from a principal component analysis of random vector $\mathbb{V}$). In addition, such a reduced-order statistical model must have the capability to represent the required mathematical properties for the random family $\{[\mathbb{A}(\mathbf{x}^1)], \ldots, [\mathbb{A}(\mathbf{x}^{N_p})]\}$. It is then proposed to introduce a family $\{[\mathbb{A}^{\text{PAPM}}(\mathbf{x}; \mathbf{s})], \mathbf{x} \in \Omega\}_{\mathbf{s}}$ of Prior Algebraic Probability Models (PAPM) to represent the matrix-valued random field $\{[\mathbb{A}(\mathbf{x})], \mathbf{x} \in \Omega\}$. For instance, the matrix-valued random field $\{[\mathbb{A}^{\text{PAPM}}(\mathbf{x})], \mathbf{x} \in \Omega\}$ can be chosen as the matrix-valued random field $\{[\mathbb{A}(\mathbf{x})] = [A^0(\mathbf{x})] + [\mathbf{A}(\mathbf{x})], \mathbf{x} \in \Omega\}$ defined in Section 8.2 in which $[A^0(\mathbf{x})]_{IJ} = C^0_{ijkh}(\mathbf{x})$. We can then deduce a family $\{\mathbb{V}^{\text{PAPM}}(\mathbf{s})\}_{\mathbf{s}}$ of PAPM for random vector $\mathbb{V}$. This family is defined on probability space $(\Theta, \mathcal{T}, \mathcal{P})$ and depends on the vector-valued parameter $\mathbf{s}$ belonging to an admissible set $\mathcal{C}_{\text{ad}}$. The knowledge of such a family means that the family $\{P^{\text{PAPM}}_{\mathbb{V}}(dv; \mathbf{s}), \mathbf{s} \in \mathcal{C}_{\text{ad}}\}$ of probability distributions on $\mathbb{R}^{m_{\mathbb{V}}}$ of the family of random vectors $\{\mathbb{V}^{\text{PAPM}}(\mathbf{s}), \mathbf{s} \in \mathcal{C}_{\text{ad}}\}$ is known. In addition, it is assumed that a generator of ν_{KL} independent realizations $\mathbb{V}^{\text{PAPM}}(\theta_1; \mathbf{s}), \ldots, \mathbb{V}^{\text{PAPM}}(\theta_{\nu_{\text{KL}}}; \mathbf{s})$, for $\theta_1, \ldots, \theta_{\nu_{\text{KL}}}$ belonging to Θ, is available. In practice, vector-valued parameter $\mathbf{s}$ will be chosen as a vector with a very low dimension. For instance, for the example of PAPM introduced in Section 8.2, the components of vector $\mathbf{s}$ can be chosen as the entries of the mean value $[\underline{a}]$, the spatial correlation lengths $L^{jj'}_k$ and the dispersion parameter δ which controls the statistical fluctuations of matrix-valued random field $\{[\mathbb{A}^{\text{PAPM}}(\mathbf{x})], \mathbf{x} \in \Omega\}$ (see the end of Section 8.2.2-(ii.1)).

Step 2. *Identification of an optimal PAPM in the constructed family using the experimental data sets*. This step consists in using the experimental data $\{\mathbf{u}^{\text{exp},1}, \ldots, \mathbf{u}^{\text{exp},\nu_{\text{exp}}}\}$ to identify the optimal value $\mathbf{s}^{\text{opt}}$ of parameter

$\mathbf{s}$. Using the computational model (see Eq. (8.2)) and the family $\mathbb{V}^{\text{PAPM}}(\mathbf{s})$ of PAPM for $\mathbb{V}$, we can construct the family $\{\mathbf{U}^{\text{obs,PAPM}}(\mathbf{s})\,,\mathbf{s}\in\mathcal{C}_{\text{ad}}\}$ of random observation vectors such that $\mathbf{U}^{\text{obs,PAPM}}(\mathbf{s}) = \mathbf{h}^{\text{obs}}(\mathbb{V}^{\text{PAPM}}(\mathbf{s}))$ for $\mathbf{s}\in\mathcal{C}_{\text{ad}}$. The optimal PAPM is then obtained in finding the optimal value $\mathbf{s}^{\text{opt}}$ of $\mathbf{s}$ which minimizes an adapted "distance" $J(\mathbf{s})$ (cost function) between the family $\{\mathbf{U}^{\text{obs,PAPM}}(\mathbf{s})\,,\mathbf{s}\in\mathcal{C}_{\text{ad}}\}$ of random observation vectors and the family of experimental data $\{\mathbf{u}^{\text{exp},1},\ldots,\mathbf{u}^{\text{exp},\nu_{\text{exp}}}\}$. We then obtain the optimal PAPM denoted by $\mathbb{V}^{\text{PAPM}} = \mathbb{V}^{\text{PAPM}}(\mathbf{s}^{\text{opt}})$. Several methods can be used to define the cost function $J(\mathbf{s})$, such as the moment method, the least-square method, the maximum likelihood method, etc. By construction of the PAPM, the dimension of vector $\mathbf{s}$ is much smaller than $\nu_{\text{exp}}\times m_{\text{obs}}$. Consequently, a method such as the maximum likelihood is not really necessary and a method in the class of the least-square method is generally sufficient and efficient (see the details in (Soize 2010b)).

Step 3. *Construction of the statistical reduced-order optimal PAPM.* For $\ell = 1,\ldots,\nu_{\text{KL}}$, let $\mathbb{V}^{\text{PAPM}}(\theta_\ell)$ be ν_{KL} independent realizations of the optimal PAPM $\mathbb{V}^{\text{PAPM}}$. The mean value $\underline{\mathbb{V}} = E\{\mathbb{V}^{\text{PAPM}}\}$ of the random vector $\mathbb{V}^{\text{PAPM}}$ and its positive-definite symmetric $(m_{\mathbb{V}}\times m_{\mathbb{V}})$ real covariance matrix $[C_{\mathbb{V}^{\text{PAPM}}}] = E\{(\mathbb{V}^{\text{PAPM}}-\underline{\mathbb{V}})\,(\mathbb{V}^{\text{PAPM}}-\underline{\mathbb{V}})^T\}$ are estimated using independent realizations $\{\mathbb{V}^{\text{PAPM}}(\theta_\ell), \ell = 1,\ldots\nu_{\text{KL}}\}$. The dominant eigenspace of the eigenvalue problem $[C_{\mathbb{V}^{\text{PAPM}}}]\,\mathbb{W}^j = \lambda_j\mathbb{W}^j$ is then constructed. Let $[\mathbb{W}] = [\mathbb{W}^1\ldots\mathbb{W}^n]$ be the $(m_{\mathbb{V}}\times n)$ real matrix of the n eigenvectors associated with the n largest eigenvalues $\lambda_1\geq\lambda_2\geq\ldots\geq\lambda_n>0$ such that $[\mathbb{W}]^T\,[\mathbb{W}] = [I_n]$, in which $[I_n]$ is the $(n\times n)$ unity matrix. The statistical reduced-order optimal PAPM is then written as

$$\mathbb{V}^{\text{PAPM}} \simeq \underline{\mathbb{V}} + \sum_{j=1}^{n}\sqrt{\lambda_j}\,\eta_j^{\text{PAPM}}\,\mathbb{W}^j\,, \tag{8.48}$$

in which $\boldsymbol{\eta}^{\text{PAPM}} = (\eta_1^{\text{PAPM}},\ldots,\eta_n^{\text{PAPM}})$ is a second-order centered random variable with values in $\mathbb{R}^n$ such that

$$E\{\boldsymbol{\eta}^{\text{PAPM}}\} = \mathbf{0}\,,\quad E\{\boldsymbol{\eta}^{\text{PAPM}}\,(\boldsymbol{\eta}^{\text{PAPM}})^T\} = [I_n]\,. \tag{8.49}$$

The mean-square convergence of the right-hand side in Eq. (8.48)) with respect to the reduced-order n is studied in constructing the error function

$$n\mapsto \text{err}(n) = 1 - \frac{\sum_{j=1}^{n}\lambda_j}{\text{tr}[C_{\mathbb{V}^{\text{PAPM}}}]}\,, \tag{8.50}$$

which is a monotonic decreasing function from $\{1,\ldots,m_{\mathbb{V}}\}$ into $[0\,,1]$ and such that $\text{err}(m_{\mathbb{V}}) = 0$. The ν_{KL} independent realizations $\boldsymbol{\eta}^{\text{PAPM}}(\theta_1)$,

$\ldots, \boldsymbol{\eta}^{\text{PAPM}}(\theta_{\nu_{\text{KL}}})$ can then be deduced from the realizations $\mathbb{V}^{\text{PAPM}}(\theta_1), \ldots, \mathbb{V}^{\text{PAPM}}(\theta_{\nu_{\text{KL}}})$ using, for $j = 1, \ldots, n$ and for $\ell = 1, \ldots \nu_{\text{KL}}$, the equation

$$\eta_j^{\text{PAPM}}(\theta_\ell) = \frac{1}{\sqrt{\lambda_j}} \, (\mathbb{V}^{\text{PAPM}}(\theta_\ell) - \underline{\mathbb{V}})^T \, \mathbb{W}^j . \tag{8.51}$$

Step 4. *Construction of the PCE with deterministic VVC of the reduced-order optimal PAPM*. This step consists in constructing an approximation $\boldsymbol{\eta}^{\text{chaos}}(N) = (\eta_1^{\text{chaos}}(N), \ldots, \eta_n^{\text{chaos}}(N))$ of $\boldsymbol{\eta}^{\text{PAPM}}$ by using a PCE, such that

$$\boldsymbol{\eta}^{\text{PAPM}} \simeq \boldsymbol{\eta}^{\text{chaos}}(N), \quad \boldsymbol{\eta}^{\text{chaos}}(N) = \sum_{\alpha=1}^{N} \mathbf{y}^\alpha \, \Psi_\alpha(\Xi), \tag{8.52}$$

in which the real valued random variables $\Psi_1(\Xi), \ldots, \Psi_N(\Xi)$ are the renumbered normalized Hermite polynomials of the $\mathbb{R}^{N_g}$-valued normalized Gaussian random variable $\Xi = (\Xi_1, \ldots, \Xi_{N_g})$ (therefore, $E\{\Xi\} = \mathbf{0}$ and $E\{\Xi \, \Xi^T\} = [I_{N_g}]$), defined on probability space $(\Theta, \mathcal{T}, \mathcal{P})$, such that for all α and β in $\{1, \ldots, N\}$,

$$E\{\Psi_\alpha(\Xi)\} = 0, \quad E\{\Psi_\alpha(\Xi) \, \Psi_\beta(\Xi)\} = \delta_{\alpha\beta}, \tag{8.53}$$

where $\delta_{\alpha\beta}$ is the Kronecker symbol. It should be noted that the constant Hermite polynomial with index $\alpha = 0$ is not included in Eq. (8.52). If N_d is the integer number representing the maximum degree of the Hermite polynomials, then the number N of chaos in Eq. (8.52) is

$$N = h(N_g, N_d) = (N_d + N_g)! \, / (N_d! \, N_g!) - 1. \tag{8.54}$$

In Eq. (8.52)), symbol "$\simeq$" means that the mean-square convergence is reached for N sufficiently large and the deterministic VVC which must be identified are the N vectors $\mathbf{y}^1, \ldots, \mathbf{y}^N$ in $\mathbb{R}^n$. Taking into account Eqs. (8.49), (8.52) and (8.53), it can be deduced that vectors $\mathbf{y}^1, \ldots, \mathbf{y}^N$ must verify the following equation,

$$\sum_{\alpha=1}^{N} \mathbf{y}^\alpha \, \mathbf{y}^{\alpha T} = [I_n], \tag{8.55}$$

and consequently, for all N, we have $E\{\boldsymbol{\eta}^{\text{chaos}}(N) \, (\boldsymbol{\eta}^{\text{chaos}}(N))^T\} = [I_n]$. In order to control the quality of the convergence of the series in Eq. (8.52) with respect to N (which is mean-square convergent), we have introduced in (Soize 2010b) an unusual L^1-log error function which allows the errors of the very small values of the probability density function (the tails of the probability density function) to be quantified. For a fixed value of N, such a quantification of the error is summarized hereinafter. Let $e \mapsto p_{\eta_j^{\text{PAPM}}}(e)$ be the probability density function of the random variable η_j^{PAPM}. For all $\mathbf{y}^1, \ldots, \mathbf{y}^N$ fixed in $\mathbb{R}^n$ and satisfying Eq. (8.55)), let $e \mapsto p_{\eta_j^{\text{chaos}}(N)}(e\,;\mathbf{y}^1, \ldots, \mathbf{y}^N)$ be the probability density function of random variable $\eta_j^{\text{chaos}}(N)$. The convergence of the sequence of random vectors $\{\boldsymbol{\eta}^{\text{chaos}}(N)\}_N$ towards $\boldsymbol{\eta}^{\text{PAPM}}$ is then controlled with the L^1-log error defined by

$$\text{err}_j(N_g, N_d) = \int_{\text{BI}_j} |\log_{10} p_{\eta_j^{\text{PAPM}}}(e) - \log_{10} p_{\eta_j^{\text{chaos}}(N)}(e\,;\mathbf{y}^1, \ldots, \mathbf{y}^N)|\, de\,, \tag{8.56}$$

in which BI_j is a bounded interval of the real line which is adapted to the problem (see the details in (Soize 2010b)). The estimation of $p_{\eta_j^{\text{PAPM}}}(e)$ is carried out using the kernel density estimation method (Bowman and Azzalini 1997) with the independent realizations $\eta_j^{\text{PAPM}}(\theta_1), \ldots, \eta_j^{\text{PAPM}}(\theta_{\nu_{\text{KL}}})$ calculated using Eq. (8.51) in Step 3. Similarly, for a given value of $\mathbf{y}^1, \ldots, \mathbf{y}^N$, the estimation of $p_{\eta_j^{\text{chaos}}(N)}(e\,;\mathbf{y}^1, \ldots, \mathbf{y}^N)$ is carried out using Eq. (8.52) and ν independent realizations $\boldsymbol{\Xi}(\theta_1), \ldots, \boldsymbol{\Xi}(\theta_\nu)$ of the normalized Gaussian vector $\boldsymbol{\Xi}$ defined on probability space $(\Theta, \mathcal{T}, \mathcal{P})$ with $\theta_1, \ldots, \theta_\nu$ in Θ. For the random vector $\boldsymbol{\eta}^{\text{chaos}}(N)$, the L^1-log error function is denoted as $\text{err}(N_g, N_d)$ and is defined by

$$\text{err}(N_g, N_d) = \frac{1}{n} \sum_{j=1}^{n} \text{err}_j(N_g, N_d). \tag{8.57}$$

It should be noted that Eqs. (8.56) and (8.57) are not used to identify $\mathbf{y}^1, \ldots, \mathbf{y}^N$, but only to evaluate, for each fixed value of N and for given $\mathbf{y}^1, \ldots, \mathbf{y}^N$, the quality of the approximation $\boldsymbol{\eta}^{\text{PAPM}} \simeq \boldsymbol{\eta}^{\text{chaos}}(N)$. For each fixed value of N, the identification of $\mathbf{y}^1, \ldots, \mathbf{y}^N$ is performed using the maximum likelihood method as done in (Desceliers et al. 2006; 2007, Das et al. 2008, Soize 2010b). Taking into account that the dependent random variables $\eta_1^{\text{chaos}}(N), \ldots, \eta_n^{\text{chaos}}(N)$ are not correlated, the following approximation $\mathcal{L}(\mathbf{y}^1, \ldots, \mathbf{y}^N)$ of the log-likelihood function is introduced

$$\mathcal{L}(\mathbf{y}^1, \ldots, \mathbf{y}^N) = \sum_{j=1}^{n} \sum_{\ell=1}^{\nu_{\text{KL}}} \log_{10} p_{\eta_j^{\text{chaos}}(N)}(\eta_j^{\text{PAPM}}(\theta_\ell)\,;\mathbf{y}^1, \ldots, \mathbf{y}^N). \tag{8.58}$$

The optimal value $(\underline{\mathbf{y}}^1, \ldots, \underline{\mathbf{y}}^N)$ of $(\mathbf{y}^1, \ldots, \mathbf{y}^N)$ is then given by

$$(\underline{\mathbf{y}}^1, \ldots, \underline{\mathbf{y}}^N) = \arg \max_{(\mathbf{y}^1, \ldots, \mathbf{y}^N) \in \mathcal{C}_{\text{ad}}^N} \mathcal{L}(\mathbf{y}^1, \ldots, \mathbf{y}^N), \tag{8.59}$$

in which $\mathcal{C}_{\text{ad}}^N$ is such that

$$\mathcal{C}_{\text{ad}}^N = \{(\mathbf{y}^1, \ldots, \mathbf{y}^N) \in (\mathbb{R}^n)^N \ , \ \sum_{\alpha=1}^{N} \mathbf{y}^\alpha \, \mathbf{y}^{\alpha\, T} = [I_n]\}. \tag{8.60}$$

For the high-dimension case, that is to say for $n \times N$ very large, solving the optimization problem defined by Eq. (8.59) with Eq. (8.60) is a very difficult problem which has been solved in the last decade only for small values of n and N. Such a challenging problem has been solved in (Soize 2010b) thanks to the introduction of two novel algorithms:

(i) The first one allows us to generate the independent realizations $\Psi_\alpha(\Xi(\theta_\ell))$ of $\Psi_\alpha(\Xi)$ with high degree N_d of the polynomials Ψ_α, for $\alpha = 1, \ldots, N$ and $\ell = 1, \ldots, \nu$. Introducing the $(\nu \times N)$ real matrix $[\Psi]$ such that $[\Psi]_{\ell\alpha} = \Psi_\alpha(\Xi(\theta_\ell))$, matrix $[\Psi]$ is computed as explained in Section 8.4 (Soize and Desceliers 2010) to preserve the orthogonality conditions defined by Eq. (8.53) for any large values of N_g and N_d (note that if an usual numerical re-orthogonalization was performed, then the independence of the realizations would be lost, that is not acceptable and consequently, another approach is then required).

(ii) The details of the second one are given in (Soize 2010b) and allows the high-dimension optimization problem defined by Eq. (8.59) with Eq. (8.60) to be solved with a reasonable CPU time, the constraint defined by Eq. (8.55) being automatically and exactly satisfied.

(iii) The random response vector $\mathbf{U}^{\text{PAPM}} = (\mathbf{U}^{\text{obs,PAPM}}, \mathbf{U}^{\text{nobs,PAPM}})$ of the computational stochastic model (see Eq. (8.2)), corresponding to the optimal PAPM represented by the PCE, is given by $\mathbf{U}^{\text{PAPM}} = \mathbf{h}(\mathbb{V}^{\text{PAPM}})$ in which $\mathbb{V}^{\text{PAPM}} \simeq \underline{\mathbb{V}} + \sum_{j=1}^{n} \sqrt{\lambda_j}\, \eta_j^{\text{PAPM}}\, \mathbb{W}^j$ with $\boldsymbol{\eta}^{\text{PAPM}} \simeq \sum_{\alpha=1}^{N} \underline{\mathbf{y}}^\alpha \, \Psi_\alpha(\Xi)$. The independent realizations $\{\mathbf{U}^{\text{PAPM}}(\theta_\ell), \ell = 1, \ldots, \nu\}$ of $\mathbf{U}^{\text{PAPM}}$ can then be calculated. For $1 \leq k \leq m_{\text{obs}}$, let U_k^{PAPM} be a component of the random observation vector $\mathbf{U}^{\text{obs,PAPM}}$ while, if $m_{\text{obs}} + 1 \leq k \leq m$, then U_k^{PAPM} represents a component of the random vector $\mathbf{U}^{\text{nobs,PAPM}}$. The probability density function $u_k \mapsto p_{U_k^{\text{PAPM}}}(u_k)$ on $\mathbb{R}$ of the random variable U_k^{PAPM} is then estimated using the above independent realizations and the kernel density estimation method (Bowman and Azzalini 1997).

8.4 Computational aspects for constructing realizations of polynomial chaos in high dimension

In this section, we present the theory developed and validated in (Soize and Desceliers 2010). The notations introduced in Section 8.3 are used.

8.4.1 *Statistically independent realizations of multivariate monomials*

Let $\boldsymbol{\Xi} = (\Xi_1, \ldots, \Xi_{N_g})$ be the $\mathbb{R}^{N_g}$-valued random vector of the independent centered random variables $\Xi_1, \ldots, \Xi_{N_g}$ for which the probability density functions (with respect to the Lebesgue measure $d\xi$ on the real line) are denoted by $p_{\Xi_1}(\xi), \ldots, p_{\Xi_{N_g}}(\xi)$. It should be noted that random vector $\boldsymbol{\Xi}$ is not assumed to be Gaussian but can follow an arbitrary probability distribution (in Section 8.3, these results are used for a Gaussian probability distribution). For all $\boldsymbol{\alpha} = (\alpha_1, \ldots, \alpha_{N_g})$ in $\mathbb{N}^{N_g}$ (including the null multi-index $(0, \ldots, 0)$) and for all $\boldsymbol{\xi} = (\xi_1, \ldots, \xi_{N_g})$ belonging to $\mathbb{R}^{N_g}$, the multivariate monomial $\mathcal{M}_{\boldsymbol{\alpha}}(\boldsymbol{\xi})$ is defined by

$$\mathcal{M}_{\boldsymbol{\alpha}}(\boldsymbol{\xi}) = \xi_1^{\alpha_1} \times \ldots \times \xi_{N_g}^{\alpha_{N_g}} . \tag{8.61}$$

Let us consider the set $\{\mathcal{M}_{\boldsymbol{\alpha}}(\boldsymbol{\xi}), \boldsymbol{\alpha}; |\boldsymbol{\alpha}| = 0, \ldots, N_d\}$ which contains $\widehat{N} = (N_d + N_g)!/\,(N_d!\,N_g!)$ multivariate monomials (that is to say, we have $\widehat{N} = N + 1$, in which N is defined by Eq. (8.54)). This set of multivariate monomials is renumbered as $\mathcal{M}_1(\boldsymbol{\xi}), \ldots, \mathcal{M}_{\widehat{N}}(\boldsymbol{\xi})$ such that the set $\{\mathcal{M}_{\boldsymbol{\alpha}}(\boldsymbol{\xi}), \boldsymbol{\alpha}; |\boldsymbol{\alpha}| = 0, \ldots, N_d\}$ of length $\widehat{N}$ is identified to the set $\{\mathcal{M}_1(\boldsymbol{\xi}), \ldots, \mathcal{M}_{\widehat{N}}(\boldsymbol{\xi})\}$. It is assumed that $\mathcal{M}_1(\boldsymbol{\xi}) = 1$ is the constant monomial. It should be noted that the random variables $\mathcal{M}_1(\boldsymbol{\Xi}), \ldots, \mathcal{M}_{\widehat{N}}(\boldsymbol{\Xi})$ are not normalized, centered or orthogonal. Let ν be an integer such that $\nu > \widehat{N}$. Let $\boldsymbol{\Xi}(\theta_1), \ldots, \boldsymbol{\Xi}(\theta_\nu)$ be ν independent realizations of random vector $\boldsymbol{\Xi}$. Then the ν independent realizations of the $\widehat{N}$ random multivariate monomials $\mathcal{M}_1(\boldsymbol{\Xi}), \ldots, \mathcal{M}_{\widehat{N}}(\boldsymbol{\Xi})$ are represented by the $\nu \times \widehat{N}$ real numbers $\{\mathcal{M}_j(\boldsymbol{\Xi}(\theta_\ell))\}_{\ell j}$.

8.4.2 *Centered statistically independent realizations of orthonormal multivariate polynomials*

We introduce the estimation $\{\underline{\mathcal{M}}_j, j = 1, \ldots, \widehat{N}\}$ of the mean values of the realizations $\{\mathcal{M}_j(\boldsymbol{\Xi}(\theta_\ell)), j = 1, \ldots, \widehat{N}\}$ such that $\underline{\mathcal{M}}_1 = 0$ and such that, for all j in $\{2, \ldots, \widehat{N}\}$, $\underline{\mathcal{M}}_j = \nu^{-1} \sum_{\ell=1}^{\nu} \mathcal{M}_j(\Xi(\theta_\ell))$. We then introduce the $(\nu \times \widehat{N})$ real matrix $[\mathcal{M}]$ of the centered realizations such that

$$[\mathcal{M}]_{\ell j} = \mathcal{M}_j(\Xi(\theta_\ell)) - \underline{\mathcal{M}}_j. \tag{8.62}$$

The main idea is to construct a $(\widehat{N} \times \widehat{N})$ real matrix $[A]$ such that the independent realizations $[\Psi]_{\ell j} = \Psi_j(\Xi(\theta_\ell))$ of the polynomial chaos $\Psi_j(\Xi)$ can be written as

$$[\Psi] = [\mathcal{M}]\,[A]. \tag{8.63}$$

With such a construction, the realizations $\{\Psi_j(\Xi(\theta_\ell)), \ell = 1, \ldots, \nu\}$ are independent, because $[\Psi]_{\ell j} = \sum_{k=1}^{\widehat{N}} [\mathcal{M}]_{\ell k}\,[A]_{kj}$ shows that the rows stay independent. In addition, due to Eq. (8.62), the estimation of the mean value of each polynomial chaos is zero except the constant polynomial chaos. It should be noted that $\{[\Psi]^T\,[\Psi]\}_{jk}/\nu$ is the estimation of $E\{\Psi_j(\Xi)\,\Psi_k(\Xi)\}$ which has to be equal to the Kronecker symbol δ_{jk}. Consequently, the orthogonality of the polynomial chaos will be preserved if $[\Psi]^T\,[\Psi]/\nu = [I_{\widehat{N}}]$ in which $[I_{\widehat{N}}]$ is the unity matrix. Substituting Eq. (8.63) into $[\Psi]^T\,[\Psi]/\nu = [I_{\widehat{N}}]$ yields

$$[A]^T\,[\mathcal{M}]^T\,[\mathcal{M}]\,[A] = \nu\,[I_{\widehat{N}}]. \tag{8.64}$$

which shows that the matrix $[A]$ is related to the singular value decomposition of matrix $[\mathcal{M}]$. Below, we detail the algorithm for the direct construction of matrix $[\Psi]$ without explicitly constructing matrix $[A]$ and then, without performing the product $[\mathcal{M}]\,[A]$.

Let $[\mathcal{M}] = [\mathbb{U}]\,[\mathcal{S}]\,[V]^T$ be the singular value decomposition of matrix $[\mathcal{M}]$ in which $[\mathbb{U}]$ is a $(\nu \times \nu)$ real unitary matrix, where $[\mathcal{S}]$ is a $(\nu \times \widehat{N})$ real matrix whose diagonal elements are nonnegative and are ordered in decreasing values, and where $[V]$ is a $(\widehat{N} \times \widehat{N})$ real unitary matrix. Since $\nu > \widehat{N}$, there are $\nu - \widehat{N}$ zero singular values that we remove hereinafter. Therefore, let $[U]$ be the $(\nu \times \widehat{N})$ real matrix whose $\widehat{N}$ columns are made up of the $\widehat{N}$ first columns of matrix $[\mathbb{U}]$. We then have

$$[U]^T\,[U] = [I_{\widehat{N}}]. \tag{8.65}$$

Let $[S]$ be the diagonal $(\widehat{N} \times \widehat{N})$ real matrix made up of the block of the non zero singular values of matrix $[\mathcal{S}]$. We can then write

$$[\mathcal{M}] = [U]\,[S]\,[V]^T. \tag{8.66}$$

Multiplying Eq. (8.66) by the invertible matrix $([S]\,[V]^T)^{-1}$ which is $[V]$ $[S]^{-1}$ yields $[\mathcal{M}]\,[V]\;[S]^{-1} = [U]$. Substituting $[U] = [\mathcal{M}]\,[V]\;[S]^{-1}$ into

Eq. (8.65) and multiplying the result by ν yield

$$(\nu^{1/2}\,[V]\,[S]^{-1})^T\,[\mathcal{M}]^T\,[\mathcal{M}]\,(\nu^{1/2}\,[V]\,[S]^{-1}) = \nu\,[I_{\widehat{N}}]. \tag{8.67}$$

Comparing with Eq. (8.64) with Eq. (8.67) yields a solution for $[A]$ which can be written as

$$[A] = \nu^{1/2}\,[V]\,[S]^{-1}. \tag{8.68}$$

Substituting Eq. (8.68) into Eq. (8.63) yields $[\Psi] = \nu^{1/2}\,[\mathcal{M}]\,[V]\,[S]^{-1}$. Finally, substituting Eq. (8.66) into the last equation yields

$$[\Psi] = \sqrt{\nu}\,[U]. \tag{8.69}$$

It should be noted that, in Eq. (8.69), matrix $[U]$ is directly constructed through the SVD of matrix $[\mathcal{M}]$. Summarizing, we have the following statistical estimations. Let j_0 be the column index such that $|[\Psi]_{\ell j_0}| = 1$ for all $\ell = 1,\ldots,\nu$ and corresponding to the ν independent realizations of the constant chaos polynomial $\Psi_{j_0}(\Xi) = 1$. We then have

$$E\{\Psi_{j_0}(\Xi)\} \simeq \frac{1}{\nu}\sum_{\ell=1}^{\nu}[\Psi]_{\ell j_0} = 1. \tag{8.70}$$

For j in $\{1,\ldots,\widehat{N}\}$ but different from j_0, we have

$$E\{\Psi_j(\Xi)\} \simeq \frac{1}{\nu}\sum_{\ell=1}^{\nu}[\Psi]_{\ell j} = 0. \tag{8.71}$$

For j and k in $\{1,\ldots,\widehat{N}\}$, we also have

$$E\{\Psi_j(\Xi)\,\Psi_k(\Xi)\} \simeq \frac{1}{\nu}\sum_{\ell=1}^{\nu}[\Psi]_{\ell j}\,[\Psi]_{\ell k} = \delta_{jk}. \tag{8.72}$$

8.5 Prior probability model of the random VVC

8.5.1 *Prior model*

Let $(\Theta',\mathcal{T}',\mathcal{P}')$ be a second probability space. Let $\mathbb{V}^{\text{prior}}$ be the prior probability model of $\mathbb{V}^{\text{PAPM}}$, defined as the $\mathbb{R}^{m_{\mathbb{V}}}$-valued random variable on the probability space $(\Theta'\times\Theta,\mathcal{T}'\otimes\mathcal{T},\mathcal{P}'\otimes\mathcal{P})$, such that

$$\mathbb{V}^{\text{prior}} = \underline{\mathbb{V}} + \sum_{j=1}^{n}\sqrt{\lambda_j}\,\eta_j^{\text{prior}}\,\mathbb{W}^j. \tag{8.73}$$

The prior probability model $\boldsymbol{\eta}^{\text{prior}} = (\eta_1^{\text{prior}}, \ldots, \eta_n^{\text{prior}})$ is a $\mathbb{R}^n$-valued random variable defined on $(\Theta' \times \Theta, \mathcal{T}' \otimes \mathcal{T}, \mathcal{P}' \otimes \mathcal{P})$, which is written as the following PCE with random VVC,

$$\boldsymbol{\eta}^{\text{prior}} = \sum_{\alpha=1}^{N} \mathbf{Y}^{\alpha,\text{prior}} \, \Psi_\alpha(\Xi). \tag{8.74}$$

The family of the $\mathbb{R}^n$-valued random variables $\{\mathbf{Y}^{1,\text{prior}}, \ldots, \mathbf{Y}^{N,\text{prior}}\}$ are defined on probability space $(\Theta', \mathcal{T}', \mathcal{P}')$. We introduce the random vector $\mathbb{Y}^{\text{prior}}$ with values in $\mathbb{R}^{Nn}$, defined on $(\Theta', \mathcal{T}', \mathcal{P}')$, such that

$$\mathbb{Y}^{\text{prior}} = (\mathbf{Y}^{1,\text{prior}}, \ldots, \mathbf{Y}^{N,\text{prior}}), \tag{8.75}$$

and for which its probability distribution is assumed to be represented by a probability density function $\mathbb{y} \mapsto p_{\mathbb{Y}}^{\text{prior}}(\mathbb{y})$ on $\mathbb{R}^{Nn}$ with respect to the measure $d\mathbb{y} = d\mathbb{y}_1 \ldots d\mathbb{y}_{Nn}$.

8.5.2 Independent realizations of the prior model

For all (θ', θ) in $\Theta' \times \Theta$, the realization $\mathbb{V}^{\text{prior}}(\theta', \theta)$ of $\mathbb{V}^{\text{prior}}$ is written as

$$\mathbb{V}^{\text{prior}}(\theta', \theta) = \underline{\mathbb{V}} + \sum_{j=1}^{n} \sqrt{\lambda_j} \, \eta_j^{\text{prior}}(\theta', \theta) \, \mathbb{W}^j, \tag{8.76}$$

in which the realization $\boldsymbol{\eta}^{\text{prior}}(\theta', \theta) = (\eta_1^{\text{prior}}(\theta', \theta), \ldots, \eta_n^{\text{prior}}(\theta', \theta))$ of $\boldsymbol{\eta}^{\text{prior}}$ is written as

$$\boldsymbol{\eta}^{\text{prior}}(\theta', \theta) = \sum_{\alpha=1}^{N} \mathbf{Y}^{\alpha,\text{prior}}(\theta') \, \Psi_\alpha(\Xi(\theta)). \tag{8.77}$$

The realization $\mathbb{Y}^{\text{prior}}(\theta')$ of $\mathbb{Y}^{\text{prior}}$ is given by

$$\mathbb{Y}^{\text{prior}}(\theta') = (\mathbf{Y}^{1,\text{prior}}(\theta'), \ldots, \mathbf{Y}^{N,\text{prior}}(\theta')). \tag{8.78}$$

8.5.3 Probability distribution of the prior model

Let $\varepsilon \geq 0$ be any given positive or null real number. The probability density function $\mathbb{y} \mapsto p_{\mathbb{Y}}^{\text{prior}}(\mathbb{y})$ on $\mathbb{R}^{nN}$ of the random vector $\mathbb{Y}^{\text{prior}} = (\mathbf{Y}^{1,\text{prior}}, \ldots, \mathbf{Y}^{N,\text{prior}})$ is chosen as follows. The random vectors $\mathbf{Y}^{1,\text{prior}}, \ldots,$ $\mathbf{Y}^{N,\text{prior}}$ are mutually independent and such that,

$$\mathbf{Y}^{\alpha,\text{prior}} = 2\varepsilon \, |\underline{\mathbf{y}}^\alpha| \, \mathbb{U}_\alpha + \underline{\mathbf{y}}^\alpha - \varepsilon \, |\underline{\mathbf{y}}^\alpha|, \tag{8.79}$$

in which $|\underline{\mathbf{y}}^\alpha|$ is the vector $(|\underline{y}_1^\alpha|, \ldots, |\underline{y}_n^\alpha|)$ where $\underline{\mathbf{y}}^1, \ldots, \underline{\mathbf{y}}^N$ are the N known vectors in $\mathbb{R}^n$ calculated with Eq. (8.59). In Eq. (8.79), $\{\mathbb{U}_1, \ldots,$

$\mathbb{U}_N\}$ is a family of independent uniform random variables on $[0,1]$, defined on $(\Theta', \mathcal{T}', \mathcal{P}')$. Consequently, the component $Y_j^{\alpha,\text{prior}}$ of $\mathbf{Y}^{\alpha,\text{prior}}$ is a uniform random variable, centered in $\underline{y}_j^\alpha$ and the support of its probability distribution is written as

$$s_j^\alpha = [\,\underline{y}_j^\alpha - \varepsilon\,|\underline{y}_j^\alpha|\,,\,\underline{y}_j^\alpha + \varepsilon\,|\underline{y}_j^\alpha|\,]. \tag{8.80}$$

It can then be deduced that $E\{\boldsymbol{\eta}^{\text{prior}}\} = 0$ and the mean values of the random VVC are such that

$$E\{\mathbf{Y}^{\alpha,\text{prior}}\} = \underline{\mathbf{y}}^\alpha, \quad \text{for } \alpha = 1,\ldots,N. \tag{8.81}$$

The statistical fluctuations of $\mathbf{Y}^{\alpha,\text{prior}}$ around the mean value $\underline{\mathbf{y}}^\alpha$ is controlled by parameter ε. If $\varepsilon = 0$, then $\mathbf{Y}^{\alpha,\text{prior}} = \underline{\mathbf{y}}^\alpha$ (deterministic case for the VVC of the PCE computed in Step 4).

8.5.4 Subset of independent realizations for the prior model

Let $\{[\mathbb{A}^{\text{prior}}(\mathbf{x}^1)],\ldots,[\mathbb{A}^{\text{prior}}(\mathbf{x}^{N_p})]\}$ be the N_p random matrices associated with the prior model $\mathbb{V}^{\text{prior}}$. In Section 8.1-(2), we have seen that the matrix-valued random field $\{[\mathbb{A}(\mathbf{x})], \mathbf{x} \in \mathcal{I} \subset \Omega\}$ must generally satisfy mathematical properties, denoted by $\mathcal{P}_{rop}$, in order that the stochastic boundary value problem be a unique stochastic solution verifying given properties (see the analysis presented in Section 8.2). By construction, for $\varepsilon = 0$ in Eq. (8.79), property $\mathcal{P}_{rop}$ is verified. However, for $\varepsilon > 0$, such a property can be not verified for certain realizations. Consequently, the rejection method is used to construct the subset of independent realizations for which $\mathcal{P}_{rop}$ is satisfied. It is interesting to precise this type of property. If we consider the ellipticity condition defined by Eq. (8.22), it is necessary and sufficient that, for all $k = 1,\ldots,N_p$ (see Section 8.2), the random matrix $[\mathbf{A}^{\text{prior}}(\mathbf{x}^k)] = [\mathbb{A}^{\text{prior}}(\mathbf{x}^k)] - [A^0]$, associated with Eq. (8.19) for the prior model, be positive definite almost surely. In such a case, property $\mathcal{P}_{rop}$ is then defined as the following: the random matrices $\{[\mathbf{A}^{\text{prior}}(\mathbf{x}^1)],\ldots,[\mathbf{A}^{\text{prior}}(\mathbf{x}^{N_p})]\}$ are positive definite almost surely.
Let ε be fixed (not equal to zero). Let $\mathbb{Y}^{\text{prior}}(\theta'_1),\ldots,\mathbb{Y}^{\text{prior}}(\theta'_{\nu'})$ be ν' independent realizations of $\mathbb{Y}^{\text{prior}}$ for $\theta'_1,\ldots,\theta'_{\nu'}$ in Θ'. Let $\Xi(\theta_1),\ldots,\Xi(\theta_\nu)$ be the ν independent realizations of Ξ (for $\theta_1,\ldots,\theta_\nu$ in Θ) introduced in Step 4 of Section 8.3. For given $\theta'_{\ell'}$ and θ_ℓ, let $\mathbb{V}^{\text{prior}}(\theta'_{\ell'},\theta_\ell)$ be the realization of $\mathbb{V}^{\text{prior}}$ and let $[\mathbb{A}^{\text{prior}}(\mathbf{x}^1;\theta'_{\ell'},\theta_\ell)],\ldots,[\mathbb{A}^{\text{prior}}(\mathbf{x}^{N_p};\theta'_{\ell'},\theta_\ell)]$ be the corresponding realizations of $[\mathbb{A}^{\text{prior}}(\mathbf{x}^1)],\ldots,[\mathbb{A}^{\text{prior}}(\mathbf{x}^{N_p})]$. Consequently, if the family $\{[\mathbb{A}^{\text{prior}}(\mathbf{x}^1;\theta'_{\ell'},\theta_\ell)],\ldots,[\mathbb{A}^{\text{prior}}(\mathbf{x}^{N_p};\theta'_{\ell'},\theta_\ell)]\}$ verifies property $\mathcal{P}_{rop}$, then realization $(\theta'_{\ell'},\theta_\ell)$ will be kept and, if not, this realization will

be rejected. For fixed $\theta'_{\ell'}$, we then introduce the subset $\{\theta_{\ell_1},\ldots,\theta_{\ell_{\widetilde{\nu}(\ell')}}\}$ of $\{\theta_1,\ldots,\theta_\nu\}$, with $\widetilde{\nu}(\ell')\leq\nu$, for which property $\mathcal{P}_{rop}$ is verified.
It should be noted that ε will arbitrarily be fixed in the context of the use of the Bayesian method to construct the posterior model. In general, more ε will be chosen large, more $\widetilde{\nu}(\ell')$ will be small. Therefore, a compromise will have to be chosen between the number $\widetilde{\nu}(\ell')$ of realizations to get convergence of the statistical estimators and a large value of ε allowing large deviations from the prior model to be generated.

8.5.5 Prior probability density functions of the responses

For all ℓ' fixed in $\{1,\ldots,\nu'\}$, the realizations $\mathbf{U}^{\text{prior}}(\theta'_{\ell'},\theta_{\ell_j})$ for $j=1,\ldots,\widetilde{\nu}(\ell')$ of the prior random vector $\mathbf{U}^{\text{prior}}=(\mathbf{U}^{\text{obs,prior}},\mathbf{U}^{\text{nobs,prior}})=\mathbf{h}(\mathbb{V}^{\text{prior}})$ are calculated with the stochastic computational model for the prior model $\mathbb{Y}^{\text{prior}}$ of $\mathbb{Y}$. For $1\leq k\leq m_{\text{obs}}$, U_k^{prior} is a component of the random observation vector $\mathbf{U}^{\text{obs,prior}}$ while, if $m_{\text{obs}}+1\leq k\leq m$, then U_k^{prior} represents a component of the random vector $\mathbf{U}^{\text{nobs,prior}}$. The probability density function $u_k\mapsto p_{U_k^{\text{prior}}}(u_k)$ on $\mathbb{R}$ of the prior random variable U_k^{prior} is then estimated using the above independent realizations and the kernel density estimation method (Bowman and Azzalini 1997).

8.6 Posterior probability model of the random VVC using the classical Bayesian approach

In this section, we present the use of the classical Bayesian approach to construct the posterior probability model $\mathbb{Y}^{\text{post}}$ of the random VVC for which the prior probability model $\mathbb{Y}^{\text{prior}}$ has been constructed in Section 8.5.

8.6.1 Conditional probability of the vector-valued random observation for given VVC

For a given vector $\mathbb{y}=(\mathbf{y}^1,\ldots,\mathbf{y}^N)$ in $\mathbb{R}^{Nn}=\mathbb{R}^n\times\ldots\times\mathbb{R}^n$, let $\mathbf{U}=(\mathbf{U}^{\text{obs}},\mathbf{U}^{\text{nobs}})$ be the random vector with values in $\mathbb{R}^m=\mathbb{R}^{m_{\text{obs}}}\times\mathbb{R}^{m_{\text{nobs}}}$, such that $\mathbf{U}=\mathbf{h}(\mathbb{V})$ (see Eq. (8.2)) in which the random vector $\mathbb{V}$ with values in $\mathbb{R}^{m_{\mathbb{V}}}$ is given by $\mathbb{V}=\underline{\mathbb{V}}+\sum_{j=1}^n\sqrt{\lambda_j}\,\eta_j\,\mathbb{W}^j$ (see Eq. (8.73)) and for which the random vector $\boldsymbol{\eta}=(\eta_1,\ldots,\eta_n)$ with values in $\mathbb{R}^n$ is given by $\boldsymbol{\eta}=\sum_{\alpha=1}^N\mathbf{y}^\alpha\,\Psi_\alpha(\Xi)$ (see Eq. (8.74)). We introduce the conditional probability density function $\mathbf{u}^{\text{obs}}\mapsto p_{\mathbf{U}^{\text{obs}}|\mathbb{Y}}(\mathbf{u}^{\text{obs}}|\mathbb{y})$ (defined on

$\mathbb{R}^{m_{\rm obs}}$ and with respect to the measure $d\mathbf{u}^{\rm obs} = du_1^{\rm obs} \ldots du_{m_{\rm obs}}^{\rm obs})$ of random observation vector $\mathbf{U}^{\rm obs}$ if $\mathbb{Y} = (\mathbf{Y}^1, \ldots, \mathbf{Y}^N)$ is equal to the given vector $\mathbb{y} = (\mathbf{y}^1, \ldots, \mathbf{y}^N)$ in $\mathbb{R}^{Nn}$. Consequently, the random observation vector $\mathbf{U}^{\rm obs} = (U_1^{\rm obs}, \ldots, U_{m_{\rm obs}}^{\rm obs})$ depends on $\mathbb{Y} = \mathbb{y}$ and the stochastic computational model allows the conditional probability density functions $\mathbf{u}^{\rm obs} \mapsto p_{\mathbf{U}^{\rm obs}|\mathbb{Y}}(\mathbf{u}^{\rm obs}|\mathbb{y})$ and $u_k^{\rm obs} \mapsto p_{U_k^{\rm obs}|\mathbb{Y}}(u_k^{\rm obs}|\mathbb{y})$ to be calculated.

8.6.2 Formulation using the Bayesian method

The posterior random vector $\mathbf{U}^{\rm post} = (\mathbf{U}^{\rm obs,post}, \mathbf{U}^{\rm nobs,post})$ with values in $\mathbb{R}^m = \mathbb{R}^{m_{\rm obs}} \times \mathbb{R}^{m_{\rm nobs}}$ is written (see Eq. (8.2)) as

$$\mathbf{U}^{\rm post} = \mathbf{h}(\mathbb{V}^{\rm post}) \; , \; \mathbf{U}^{\rm obs,post} = \mathbf{h}^{\rm obs}(\mathbb{V}^{\rm post}) \; , \; \mathbf{U}^{\rm nobs,post} = \mathbf{h}^{\rm nobs}(\mathbb{V}^{\rm post}) \; , \tag{8.82}$$

in which the $\mathbb{R}^{m_{\mathbb{V}}}$-valued random vector $\mathbb{V}^{\rm post}$ is the posterior model of $\mathbb{V}^{\rm prior}$. Taking into account Eqs. (8.73) to (8.75), the posterior model of $\mathbb{V}^{\rm prior}$ is written as

$$\mathbb{V}^{\rm post} = \underline{\mathbb{V}} + \sum_{j=1}^{n} \sqrt{\lambda_j}\, \eta_j^{\rm post}\, \mathbb{W}^j , \tag{8.83}$$

in which the posterior probability model $\boldsymbol{\eta}^{\rm post} = (\eta_1^{\rm post}, \ldots, \eta_n^{\rm post})$ is a $\mathbb{R}^n$-valued random variable defined on $(\Theta' \times \Theta, \mathcal{T}' \otimes \mathcal{T}, \mathcal{P}' \otimes \mathcal{P})$, such that

$$\boldsymbol{\eta}^{\rm post} = \sum_{\alpha=1}^{N} \mathbf{Y}^{\alpha,\rm post}\, \Psi_\alpha(\Xi), \tag{8.84}$$

in which the family of $\mathbb{R}^n$-valued random variables $\{\mathbf{Y}^{1,\rm post}, \ldots, \mathbf{Y}^{N,\rm post}\}$ are defined on the probability space $(\Theta', \mathcal{T}', \mathcal{P}')$. As previously, we introduce the $\mathbb{R}^{Nn}$-valued random vector $\mathbb{Y}^{\rm post}$ defined on $(\Theta', \mathcal{T}', \mathcal{P}')$ such that

$$\mathbb{Y}^{\rm post} = (\mathbf{Y}^{1,\rm post}, \ldots, \mathbf{Y}^{N,\rm post}), \tag{8.85}$$

whose its probability distribution is represented by the probability density function $\mathbb{y} \mapsto p_{\mathbb{Y}}^{\rm post}(\mathbb{y})$ on $\mathbb{R}^{Nn}$ with respect to the measure $d\mathbb{y} = d\mathbb{y}_1 \ldots d\mathbb{y}_{nm}$.

Let $\mathbf{u}^{\exp,1}, \ldots, \mathbf{u}^{\exp,\nu_{\exp}}$ be the $\nu_{\exp}$ independent experimental data $\mathbf{U}^{\exp}$ corresponding to observation vector $\mathbf{U}^{\rm obs}$. The Bayesian method allows the posterior probability density function $p_{\mathbb{Y}}^{\rm post}(\mathbb{y})$ to be estimated by

$$p_{\mathbb{Y}}^{\rm post}(\mathbb{y}) = L^{\rm bayes}(\mathbb{y})\, p_{\mathbb{Y}}^{\rm prior}(\mathbb{y}), \tag{8.86}$$

in which $\mathbb{y} \mapsto L^{\text{bayes}}(\mathbb{y})$ is the likelihood function defined on $\mathbb{R}^{Nn}$, with values in $\mathbb{R}^+$, such that

$$L^{\text{bayes}}(\mathbb{y}) = \frac{\Pi_{\ell=1}^{\nu_{\text{exp}}} p_{\mathbf{U}^{\text{obs}}|\mathbb{Y}}(\mathbf{u}^{\text{exp},\ell}|\mathbb{y})}{E\{\Pi_{\ell=1}^{\nu_{\text{exp}}} p_{\mathbf{U}^{\text{obs}}|\mathbb{Y}}(\mathbf{u}^{\text{exp},\ell}|\mathbb{Y}^{\text{prior}})\}}. \tag{8.87}$$

In Eq. (8.87), $p_{\mathbf{U}^{\text{obs}}|\mathbb{Y}}(\mathbf{u}^{\text{exp},\ell}|\mathbb{y})$ is the experimental value of the conditional probability density function $p_{\mathbf{U}^{\text{obs}}|\mathbb{Y}}(\mathbf{u}^{\text{obs}}|\mathbb{y})$. Equation (8.87) shows that likelihood function L^{bayes} must verify the following equation,

$$E\{L^{\text{bayes}}(\mathbb{Y}^{\text{prior}})\} = \int_{\mathbb{R}^{Nn}} L^{\text{bayes}}(\mathbb{y})\, p_{\mathbb{Y}}^{\text{prior}}(\mathbb{y})\, d\mathbb{y} = 1. \tag{8.88}$$

8.6.3 Posterior probability density functions of the responses

The probability density function $\mathbf{u} \mapsto p_{\mathbf{U}^{\text{post}}}(\mathbf{u})$ on $\mathbb{R}^m$ of the posterior random vector $\mathbf{U}^{\text{post}}$ is then given by $p_{\mathbf{U}^{\text{post}}}(\mathbf{u}) = \int_{\mathbb{R}^{Nn}} p_{\mathbf{U}|\mathbb{Y}}(\mathbf{u}|\mathbb{y})\, p_{\mathbb{Y}}^{\text{post}}(\mathbb{y})\, d\mathbb{y}$ in which $p_{\mathbf{U}|\mathbb{Y}}(\mathbf{u}|\mathbb{y})$ is the conditional probability density function of $\mathbf{U}$ given $\mathbb{Y} = \mathbb{y}$ and which is constructed using the stochastic computational model defined in Section 8.1. Using Eq. (8.86), this last equation can be rewritten as

$$p_{\mathbf{U}^{\text{post}}}(\mathbf{u}) = E\{L^{\text{bayes}}(\mathbb{Y}^{\text{prior}})\, p_{\mathbf{U}|\mathbb{Y}}(\mathbf{u}|\mathbb{Y}^{\text{prior}})\}.$$

Let U_k^{post} be any component of random vector $\mathbf{U}^{\text{post}}$. For $1 \leq k \leq m_{\text{obs}}$, U_k^{post} represents a component of random observation vector $\mathbf{U}^{\text{obs,post}}$ while, if $m_{\text{obs}} + 1 \leq k \leq m$, then U_k^{post} represents a component of random vector $\mathbf{U}^{\text{nobs,post}}$. Consequently, the probability density function $u_k \mapsto p_{U_k^{\text{post}}}(u_k)$ on $\mathbb{R}$ of the posterior random variable U_k^{post} is then given by

$$p_{U_k^{\text{post}}}(u_k) = E\{L^{\text{bayes}}(\mathbb{Y}^{\text{prior}})\, p_{U_k|\mathbb{Y}}(u_k|\mathbb{Y}^{\text{prior}})\}, \tag{8.89}$$

in which $p_{U_k|\mathbb{Y}}(u_k|\mathbb{y})$ is the conditional probability density function of the real valued random variable U_k given $\mathbb{Y} = \mathbb{y}$ and which is constructed using the stochastic computational model defined in Section 8.1.

8.6.4 Computational aspects

We use the notation introduced in Section 8.5.4 concerning the realizations of $\mathbb{Y}$ and Ξ. For ν' sufficiently large, the right-hand side of Eq. (8.89) can be estimated by

$$p_{U_k^{\text{post}}}(u_k) \simeq \frac{1}{\nu'} \sum_{\ell'=1}^{\nu'} L^{\text{bayes}}(\mathbb{Y}^{\text{prior}}(\theta'_{\ell'}))\, p_{U_k|\mathbb{Y}}(u_k|\mathbb{Y}^{\text{prior}}(\theta'_{\ell'})). \tag{8.90}$$

For fixed $\theta'_{\ell'}$, the computational model defined in Section 8.1 is used to calculate the $\widetilde{\nu}(\ell')$ realizations $\mathbf{U}(\theta_{\ell_1}|\mathbb{Y}^{\rm prior}(\theta'_{\ell'})),\ldots,\mathbf{U}(\theta_{\ell_{\widetilde{\nu}(\ell')}}|\mathbb{Y}^{\rm prior}(\theta'_{\ell'}))$ for $\mathbb{y}=\mathbb{Y}^{\rm prior}(\theta'_{\ell'})$. We can then deduce

$$\mathbf{U}^{\rm obs}(\theta_{\ell_1}|\mathbb{Y}^{\rm prior}(\theta'_{\ell'})),\ldots,\mathbf{U}^{\rm obs}(\theta_{\ell_{\widetilde{\nu}(\ell')}}|\mathbb{Y}^{\rm prior}(\theta'_{\ell'})),$$

and, for all fixed k,

$$U_k(\theta_{\ell_1}|\mathbb{Y}^{\rm prior}(\theta'_{\ell'})),\ldots,U_k(\theta_{\ell_{\widetilde{\nu}(\ell')}}|\mathbb{Y}^{\rm prior}(\theta'_{\ell'})).$$

(1) Using these independent realizations and the multivariate Gaussian kernel density estimation (see Section 8.6.5), $p_{\mathbf{U}^{\rm obs}|\mathbb{Y}}(\mathbf{u}^{{\rm exp},\ell}|\mathbb{Y}^{\rm prior}(\theta'_{\ell'}))$ can be estimated and then, using Eq. (8.87), we can compute $L^{\rm bayes}(\mathbb{Y}^{\rm prior}(\theta'_{\ell'}))$ for $\ell'=1,\ldots,\nu'$.
(2) For all fixed k, using the above independent realizations and the kernel estimation method (Bowman and Azzalini 1997), we can estimate $p_{U_k|\mathbb{Y}}(u_k|\mathbb{Y}^{\rm prior}(\theta'_{\ell'}))$ and then, using Eq. (8.90), we can estimate $p_{U_k^{\rm post}}(u_k)$.

8.6.5 Estimation of a probability density function on $\mathbb{R}^m$ using the multivariate Gaussian kernel density estimation

In this section, we summarize the nonparametric estimation of a multivariate probability density function using the multivariate Gaussian kernel density estimation (see for instance (Bowman and Azzalini 1997, Terrell and Scott 1992)). Such a method is used to estimate $p_{\mathbf{U}^{\rm obs}|\mathbb{Y}}(\mathbf{u}^{{\rm exp},\ell}|\mathbb{y})$ (see Eq. (8.87)) for given $\mathbf{u}^{{\rm exp},\ell}$ and for given $\mathbb{y}$.

Let $\mathbf{S}=(S_1,\ldots,S_m)$, with $m>1$, be any second-order random variable defined on $(\Theta,\mathcal{T},\mathcal{P})$ with values in $\mathbb{R}^m$ (the components are statistically dependent and $\mathbf{S}$ is not a Gaussian random vector). Let $\mathbf{S}(\theta_1),\ldots,\mathbf{S}(\theta_\nu)$ be ν independent realizations of $\mathbf{S}$ with $\theta_1,\ldots,\theta_\nu$ in Θ. Let $P_{\mathbf{S}}(d\mathbf{s})=p_{\mathbf{S}}(\mathbf{s})\,d\mathbf{s}$ be the probability distribution defined by an unknown probability density function $\mathbf{s}\mapsto p_{\mathbf{S}}(\mathbf{s})$ on $\mathbb{R}^m$, with respect to the Lebesgue measure $d\mathbf{s}$ on $\mathbb{R}^m$. For $\mathbf{s}^0$ fixed in $\mathbb{R}^m$ and for ν sufficiently large, the multivariate kernel density estimation allows the nonparametric estimation $\widehat{p}_{\mathbf{S}}(\mathbf{s}^0)$ of $p_{\mathbf{S}}(\mathbf{s}^0)$ to be carried out using the ν independent realizations $\mathbf{S}(\theta_1),\ldots,\mathbf{S}(\theta_\nu)$. The first step consists in performing a rotation and a normalization of data in the principal component axes. The second step is devoted to the Gaussian kernel density for each direction.

Let $\widehat{\underline{\mathbf{S}}}$ and $[\widehat{C}_\mathbf{S}]$ be the usual statistical estimations of the mean value and the covariance matrix of the random vector $\mathbf{S}$ using the ν realizations. For instance, $\widehat{\underline{\mathbf{S}}} = \nu^{-1}\sum_{\ell=1}^{\nu}\mathbf{S}(\theta_\ell)$. It is assumed that $[\widehat{C}_\mathbf{S}]$ is a positive-definite symmetric $(m \times m)$ real matrix. Consequently, there is an orthogonal $(m\times m)$ real matrix $[\Phi]$ (that is to say $[\Phi]\,[\Phi]^T = [\Phi]^T\,[\Phi] = [I_m]$) such that $[\widehat{C}_\mathbf{S}]\,[\Phi] = [\Phi]\,[\lambda]$ in which $[\lambda]$ is the diagonal matrix of the positive eigenvalues. Let $\mathbf{Q} = (Q_1,\ldots,Q_m)$ be the random vector such that

$$\mathbf{S} = \widehat{\underline{\mathbf{S}}} + [\Phi]\,\mathbf{Q}, \quad \mathbf{Q} = [\Phi]^T\,(\mathbf{S} - \widehat{\underline{\mathbf{S}}}). \tag{8.91}$$

We have $p_\mathbf{S}(\mathbf{s})\,d\mathbf{s} = p_\mathbf{Q}(\mathbf{q})\,d\mathbf{q}$ and since $|\det[\Phi]| = 1$, we have $d\mathbf{s} = d\mathbf{q}$. Consequently, if we introduce $\mathbf{q}^0 = [\Phi]^T\,(\mathbf{s}^0 - \widehat{\underline{\mathbf{S}}})$, then $p_\mathbf{S}(\mathbf{s}^0) = p_\mathbf{Q}(\mathbf{q}^0)$ and therefore, the nonparametric estimation $\widehat{p}_\mathbf{S}(\mathbf{s}^0)$ of $p_\mathbf{S}(\mathbf{s}^0)$ is equal to the nonparametric estimation $\widehat{p}_\mathbf{Q}(\mathbf{q}^0)$ of $p_\mathbf{Q}(\mathbf{q}^0)$, that is to say,

$$\widehat{p}_\mathbf{S}(\mathbf{s}^0) = \widehat{p}_\mathbf{Q}(\mathbf{q}^0). \tag{8.92}$$

Using Eq. (8.91), the realizations $\mathbf{S}(\theta_1),\ldots,\mathbf{S}(\theta_\nu)$ are transformed in the realizations $\mathbf{Q}(\theta_1),\ldots,\mathbf{Q}(\theta_\nu)$ of random vector $\mathbf{Q}$ such that, for all ℓ in $\{1,\ldots,\nu\}$, $\mathbf{Q}(\theta_\ell) = [\Phi]^T\,(\mathbf{S}(\theta_\ell) - \widehat{\underline{\mathbf{S}}})$. Equation (8.92) shows that the initial problem is equivalent to the construction of the nonparametric estimation $\widehat{p}_\mathbf{Q}(\mathbf{q}^0)$ of $p_\mathbf{Q}(\mathbf{q}^0)$ using the realizations $\mathbf{Q}(\theta_1),\ldots,\mathbf{Q}(\theta_\nu)$ of random vector $\mathbf{Q}$. Let $\widehat{\underline{\mathbf{Q}}}$ and $[\widehat{C}_\mathbf{Q}]$ be the usual statistical estimations of the mean value and the covariance matrix of the random vector $\mathbf{Q}$ using the ν independent realizations $\mathbf{Q}(\theta_1),\ldots,\mathbf{Q}(\theta_\nu)$. It can be seen that

$$\widehat{\underline{\mathbf{Q}}} = 0, \quad [\widehat{C}_\mathbf{Q}] = [\lambda]. \tag{8.93}$$

The second step consists in calculating $\widehat{p}_\mathbf{Q}(\mathbf{q}^0)$ using the the multivariate kernel density estimation which is written as

$$\widehat{p}_\mathbf{Q}(\mathbf{q}^0) = \frac{1}{\nu}\sum_{\ell=1}^{\nu}\Pi_{k=1}^{m}\left\{\frac{1}{h_k}K\left(\frac{Q_k(\theta_\ell) - q_k^0}{h_k}\right)\right\}, \tag{8.94}$$

in which $h_1,\ldots,h_m$ are the smoothing parameters, $\mathbf{q}^0 = (q_1^0,\ldots,q_m^0)$ and where K is the kernel. For the multivariate Gaussian kernel density estimation, we have

$$h_k = \sqrt{[\lambda]_{kk}}\left\{\frac{4}{\nu(2+m)}\right\}^{\frac{1}{4+m}}, \quad K(v) = \frac{1}{\sqrt{2\pi}}e^{-\frac{v^2}{2}}. \tag{8.95}$$

8.7 Posterior probability model of the random VVC using a new approach derived from the Bayesian approach

In Section 8.6, we have presented the classical Bayesian approach to construct the posterior model $\mathbb{Y}^{\text{post}}$ of the prior model $\mathbb{Y}^{\text{prior}}$ of the random VVC. Nevertheless, for a very high-dimension problem (the random vector $\mathbb{Y}^{\text{post}}$ can have several millions of components), the usual Bayesian method can be improved to get a more efficient method derived from the classical one and presented below. This new approach has been introduced in (Soize 2011).

It should be noted that Eq. (8.89) can be rewritten as

$$p_{U_k^{\text{post}}}(u_k) = E\{L^{\text{lls}}\, p_{U_k|\mathbb{Y}}(u_k|\mathbb{Y}^{\text{prior}})\}, \tag{8.96}$$

in which the positive-valued random variable L^{lls}, defined on $(\Theta', \mathcal{T}', \mathcal{P}')$, is such that $L^{\text{lls}} = L^{\text{bayes}}(\mathbb{Y}^{\text{prior}})$ and such that $E\{L^{\text{lls}}\} = 1$ (see Eq. (8.88)). The ν' independent realizations of L^{lls} are $L^{\text{lls}}(\theta'_1), \ldots, L^{\text{lls}}(\theta'_{\nu'})$ such that $L^{\text{lls}}(\theta'_{\ell'}) = L^{\text{bayes}}(\mathbb{Y}^{\text{prior}}(\theta'_{\ell'}))$. With such a notation, Eq. (8.90) can be rewritten as

$$p_{U_k^{\text{post}}}(u_k) \simeq \frac{1}{\nu'} \sum_{\ell'=1}^{\nu'} L^{\text{lls}}(\theta'_{\ell'})\, p_{U_k|\mathbb{Y}}(u_k|\mathbb{Y}^{\text{prior}}(\theta'_{\ell'})), \tag{8.97}$$

and $E\{L^{\text{lls}}\} = 1$ yields

$$\frac{1}{\nu'} \sum_{\ell'=1}^{\nu'} L^{\text{lls}}(\theta'_{\ell'}) \simeq 1. \tag{8.98}$$

The method proposed consists in using Eq. (8.97), but in replacing the realizations $L^{\text{lls}}(\theta'_1), \ldots, L^{\text{lls}}(\theta'_{\nu'})$ of the random variable $L^{\text{lls}} = L^{\text{bayes}}(\mathbb{Y}^{\text{prior}})$ by the realizations $L^{\text{lls}}(\theta'_1), \ldots, L^{\text{lls}}(\theta'_{\nu'})$ of another random variable L^{lls} for which the vector $\mathbf{L}^{\text{lls}} = (L^{\text{lls}}(\theta'_1), \ldots, L^{\text{lls}}(\theta'_{\nu'}))$ is constructed as the unique solution of the following linear least-square optimization problem with nonnegativity constraints,

$$\mathbf{L}^{\text{lls}} = \arg\min_{\mathbf{L}\in\mathcal{G}_{\text{ad}}} G(\mathbf{L}). \tag{8.99}$$

The admissible set $\mathcal{G}_{\text{ad}}$ is defined as

$$\mathcal{G}_{\text{ad}} = \{\mathbf{L} = (L_1, \ldots, L_{\nu'}) \in \mathbb{R}^{\nu'};\, L_1 \geq 0, \ldots, L_{\nu'} \geq 0\,;\, \frac{1}{\nu'} \sum_{\ell'=1}^{\nu'} L_{\ell'} = 1\}. \tag{8.100}$$

The cost function $\mathbf{L} \mapsto G(\mathbf{L})$ is defined as

$$G(\mathbf{L}) = \sum_{k=1}^{m_{\text{obs}}} \int_{\mathbb{R}} \Big(\widehat{p}_{U_k^{\text{exp}}}(u_k^{\text{obs}}) - \frac{1}{\nu'} \sum_{\ell'=1}^{\nu'} L_{\ell'}\, p_{U_k^{\text{obs}}|\mathbb{Y}}(u_k^{\text{obs}}|\mathbb{Y}^{\text{prior}}(\theta'_{\ell'}))\Big)^2 \, du_k^{\text{obs}}, \tag{8.101}$$

in which $u_k^{\text{obs}} \mapsto \widehat{p}_{U_k^{\text{exp}}}(u_k^{\text{obs}})$ is an estimation of the probability density function of the random variable U_k^{exp} carried out with the experimental data $u_k^{\text{exp},1}, \ldots, u_k^{\text{exp},\nu_{\text{exp}}}$ (see Section 8.1-(3)) and using the kernel estimation method. In Eq. (8.101), the quantity $p_{U_k^{\text{obs}}|\mathbb{Y}}(u_k^{\text{obs}}|\mathbb{Y}^{\text{prior}}(\theta'_{\ell'}))$ is estimated as explained in Section 8.6.4. The optimization problem defined by Eq. (8.99) can be solved, for instance, using the algorithm described in (Lawson and Hanson 1974). The quality assessment is performed using Eq. (8.97) for k such that $m_{\text{obs}}+1 \leq k \leq m$, that is to say, when U_k^{post} represents a component of the random vector $\mathbf{U}^{\text{nobs,post}}$ which is not observed and which is then not used in Eq. (8.99) for the calculation of $\mathbf{L}^{\text{lls}}$.

In theory, the Bayesian approach, presented in Section 8.6, can be used in high dimension and for a few experimental data (small value of ν_{exp}), but in practice, for the high-dimension case (very large value of the product Nn, such as several millions), the posterior probability model significantly improves the prior model only if many experimental data are available (large value of ν_{exp}). On the other hand, the method proposed in Section 8.7 requires the estimation of the probability density function of the experimental observations U_k^{exp} (using the kernel density estimation in the context of nonparametric statistics). Such an estimation is not correct if ν_{exp} is too small and must be sufficiently large (for instance $\nu_{\text{exp}} \simeq 100$). In (Soize 2011), for an application with $Nn = 10\,625 \times 550 = 5\,843\,750$ (high-dimension case), $m_{\text{obs}} = 50$ while $m = 1\,017$ (partial data) and $\nu_{\text{exp}} = 200$ (limited data), it is proven that the posterior model constructed with the method proposed in Section 8.7 is more efficient than the Bayesian method presented in Section 8.6. Finally, it should be noted that, as soon as the independent realizations $\mathbb{Y}^{\text{prior}}(\theta'_1), \ldots, \mathbb{Y}^{\text{prior}}(\theta'_{\nu'})$ are given, Eqs. (8.99) to (8.101) correspond to the generator of random variable L^{lls} allowing the realizations $L^{\text{lls}}(\theta'_1), \ldots,$ $L^{\text{lls}}(\theta'_{\nu'})$ to be generated. For any ν' and for any realizations of $\mathbb{Y}^{\text{prior}}$, we then have a generator of realizations of L^{lls}. For any given measurable mapping $\mathbf{g}$ defined on $\mathbb{R}^m$, we can then compute the quantity

$$E\{\mathbf{g}(\mathbf{U}^{\text{post}})\} \simeq \frac{1}{\nu'} \sum_{\ell'=1}^{\nu'} \frac{L^{\text{lls}}(\theta'_{\ell'})}{\widetilde{\nu}(\ell')} \sum_{j=1}^{\widetilde{\nu}(\ell')} \mathbf{g}(\mathbf{U}^{\text{prior}}(\theta'_{\ell'}, \theta_{\ell_j})), \tag{8.102}$$

where $\mathbf{U}^{\text{prior}}(\theta'_{\ell'}, \theta_{\ell_j}) = \mathbf{h}(\mathbb{V}^{\text{prior}}(\theta'_{\ell'}, \theta_{\ell_j}))$ is calculated with the computational model and where

$$\mathbb{V}^{\text{prior}}(\theta'_{\ell'}, \theta_{\ell_j})) = \underline{\mathbb{V}} + \sum_{j=1}^{n} \sqrt{\lambda_j}\, \eta_j^{\text{prior}}(\theta'_{\ell'}, \theta_{\ell_j})\, \mathbb{W}^j , \tag{8.103}$$

with

$$\boldsymbol{\eta}^{\text{prior}}(\theta'_{\ell'}, \theta_{\ell_j}) = \sum_{\alpha=1}^{N} \mathbf{Y}^{\alpha,\text{prior}}(\theta'_{\ell'})\, \Psi_\alpha(\boldsymbol{\Xi}(\theta_{\ell_j})). \tag{8.104}$$

8.8 Comments about the applications concerning the identification of polynomial chaos expansions of random fields using experimental data

It should be noted that developments of prior algebraic probability models of non-Gaussian tensor-valued random fields can be found in (Das and Ghanem 2009, Guilleminot et al. 2009, Guilleminot and Soize 2010, Guilleminot et al. 2011, Guilleminot and Soize 2011, Soize 2006; 2008b, Ta et al. 2010). Polynomial chaos representations of experimental data can be found, for instance, in (Das et al. 2008; 2009, Desceliers et al. 2006; 2007, Guilleminot et al. 2008). Identification of Bayesian posteriors can be found in (Arnst et al. 2010) and for the high-dimension case (object of the present Section 8) in (Soize 2010b; 2011).

Chapter 9

Conclusion

We have presented an ensemble of main concepts, methodologies and results allowing important problems to be addressed in two domains.

The first one concerns the stochastic modeling of uncertainties, in particular in presence of model uncertainties induced by modeling errors in computational models. The approach proposed is based on the use of the random matrix theory allowing modeling errors to be directly taken into account at the operator levels of the boundary value problems. The fundamental aspects related to the identification of the prior and the posterior probability models using experimental data are presented. We have limited the presentation of this first domain to the context of representative linear and nonlinear dynamical and structural-acoustical problems. Such a probability methodology can be used for elliptic (static) problems and parabolic (diffusion) problems, some works being already published or in progress in this field.

The second one concerns the very difficult problem related to the identification of high-dimension polynomial chaos expansions with random coefficients for non-Gaussian tensor-valued random fields using experimental data through a boundary value problem. The methodology presented is developed in the context of mesoscale stochastic modeling of heterogeneous elastic microstructures but can be applied to many mechanical problems requiring stochastic models in high-dimension, and is particularly well adapted when the available experimental data are partial and limited. The main idea is to use a polynomial chaos expansion with random coefficients to represent the unknown non-Gaussian tensor-valued random field. The posterior probability model is constructed with the Bayes method. However, taking into account the fact that there is a very large number (due to the high-dimension aspects) of dependent random variables for which the probability distribution must be identified, it is absolutely necessary to use a prior probability model of very high quality which integrates the maximum of *a priori* available information in order that the posterior probability model be identified in a neighborhood of a reasonable *a priori* solution fitting as well as possible the experimental data.

All the presented methodologies and presented formulations have been validated with simulate experiments and/or with experimental data and we have referred the readers to papers in which the details can be found.

References

Anderson, T. W. (1958) *Introduction to Multivariate Statistical Analysis* John Wiley & Sons, New York.

Arnst, M., Clouteau, D., and Bonnet, M. (2008) Inversion of probabilistic structural models using measured transfer functions *Computer Methods in Applied Mechanics and Engineering,* 197(6-8):589–608.

Arnst, M., Clouteau, D., Chebli, H., Othman, R., and Degrande, G. (2006) A nonparametric probabilistic model for ground-borne vibrations in buildings *Probabilistic Engineering Mechanics,* 21(1):18–34.

Arnst, M. and Ghanem, R. (2008) Probabilistic equivalence and stochastic model reduction in multiscale analysis *Computer Methods in Applied Mechanics and Engineering,* 197:3584–3592.

Arnst, M., Ghanem, R., and Soize, C. (2010) Identification of bayesian posteriors for coefficients of chaos expansions *Journal of Computational Physics,* 229(9):3134–3154.

Au, S. and Beck, J. (2003a) Important sampling in high dimensions *Structural Safety,* 25(2):139–163.

Au, S. and Beck, J. (2003b) Subset simulation and its application to seismic risk based on dynamic analysis *Journal of Engineering Mechanics - ASCE,* 129(8):901–917.

Azeez, M. and Vakakis, A. (2001) Proper orthogonal decomposition (pod) of a class of vibroimpact oscillations *Journal of Sound and Vibration,* 240(5):859–889.

Babuska, I., Nobile, F., and Tempone, R. (2007) A stochastic collocation method for elliptic partial differential equations with random input data *SIAM Journal on Numerical Analysis,* 45(3):1005–1034.

Babuska, I., Tempone, R., and Zouraris, G. E. (2005) Solving elliptic boundary value problems with uncertain coefficients by the finite element method: the stochastic formulation *Computer Methods in Applied Mechanics and Engineering,* 194(12-16):1251–1294.

Batou, A. and Soize, C. (2009a) Experimental identification of turbulent fluid forces applied to fuel assemblies using an uncertain model and fretting-wear estimation *Mechanical Systems and Signal Processing*, 23(7):2141–2153.

Batou, A. and Soize, C. (2009b) Identification of stochastic loads applied to a non-linear dynamical system using an uncertain computational model and experimental responses *Computational Mechanics*, 43(4):559–571.

Batou, A., Soize, C., and Corus, M. (2011) Experimental identification of an uncertain computational dynamical model representing a family of structures *Computer and Structures*, 89(13-14):1440–1448.

Beck, J. (2010) Bayesian system identification based on probability logic *Structural Control and Health Monitoring*, 17(7):825–847.

Beck, J. and Au, S. (2002) Bayesian updating of structural models and reliability using markov chain monte carlo simulation *Journal of Engineering Mechanics - ASCE*, 128(4):380–391.

Beck, J., Chan, E., A.Irfanoglu, and et al (1999) Multi-criteria optimal structural design under uncertainty *Earthquake Engineering and Structural Dynamics*, 28(7):741–761.

Beck, J. L. and Katafygiotis, L. S. (1998) Updating models and their uncertainties. i: Bayesian statistical framework *Journal of Engineering Mechanics*, 124(4):455–461.

Bernardo, J. M. and Smith, A. F. M. (2000) *Bayesian Theory* John Wiley & Sons, Chichester.

Berveiller, M., Sudret, B., and Lemaire, M. (2006) Stochastic finite element: a non-intrusive approach by regression *European Journal of Computational Mechanics*, 15:81–92.

Blatman, G. and Sudret, B. (2007) Sparse polynomial chaos expansions and adaptive stochastic finite elements using a regression approach *Comptes Rendus Mcanique*, 336(6):518–523.

Bowman, A. W. and Azzalini, A. (1997) *Applied Smoothing Techniques for Data Analysis* Oxford University Press, Oxford.

Capiez-Lernout, E., Pellissetti, M., Pradlwarter, H., Schueller, G. I., and Soize, C. (2006) Data and model uncertainties in complex aerospace engineering systems *Journal of Sound and Vibration*, 295(3-5):923–938.

Capiez-Lernout, E. and Soize, C. (2004) Nonparametric modeling of random uncertainties for dynamic response of mistuned bladed disks *Journal of Engineering for Gas Turbines and Power*, 126(3):600–618.

Capiez-Lernout, E. and Soize, C. (2008a) Design optimization with an uncertain vibroacoustic model *Journal of Vibration and Acoustics*, 130(2):021001–1 – 021001–8.

Capiez-Lernout, E. and Soize, C. (2008b) Robust design optimization in computational mechanics *Journal of Applied Mechanics - Transactions of the ASME*, 75(2):021001–1 – 021001–11.

Capiez-Lernout, E. and Soize, C. (2008c) Robust updating of uncertain damping models in structural dynamics for low- and medium-frequency ranges *Mechanical Systems and Signal Processing*, 22(8):1774–1792.

Capiez-Lernout, E., Soize, C., Lombard, J.-P., Dupont, C., and Seinturier, E. (2005) Blade manufacturing tolerances definition for a mistuned industrial bladed disk *Journal of Engineering for Gas Turbines and Power*, 127(3):621–628.

Carlin, B. P. and Louis, T. A. (2009) *Bayesian Methods for Data Analysis* Third Edition, Chapman & Hall / CRC Press, Boca Raton.

Casella, G. and George, E. (1992) Explaining the gibbs sampler *The American Statistician*, 46(3):167–174.

Cataldo, E., Soize, C., Sampaio, R., and Desceliers, C. (2009) Probabilistic modeling of a nonlinear dynamical system used for producing voice *Computational Mechanics*, 43(2):265–275.

Chebli, H. and Soize, C. (2004) Experimental validation of a nonparametric probabilistic model of non homogeneous uncertainties for dynamical systems *Journal of the Acoustical Society of America*, 115(2):697–705.

Chen, C., Duhamel, D., and Soize, C. (2006) Probabilistic approach for model and data uncertainties and its experimental identification in structural dynamics: Case of composite sandwich panels *Journal of Sound and Vibration*, 294(1-2):64–81.

Cheung, S. and Beck, J. (2009) Bayesian model updating using hybrid monte carlo simulation with application to structural dynamic models with many uncertain parameters *Journal of Engineering Mechanics - ASCE*, 135(4):243–255.

Cheung, S. and Beck, J. (2010) Calculation of posterior probabilities for bayesian model class assessment and averaging from posterior samples based on dynamic system data *Computer-Aided Civil and Infrastructure Engineering*, 25(5):304–321.

Ching, J., Beck, J., and Porter, K. (2006) Bayesian state and parameter estimation of uncertain dynamical systems *Probabilistic Engineering Mechanics*, 21(1):81–96.

Congdon, P. (2007) *Bayesian Statistical Modelling* Second Edition, John Wiley & Sons, Chichester.

Cottereau, R., Clouteau, D., and Soize, C. (2007) Construction of a probabilistic model for impedance matrices *Computer Methods in Applied Mechanics and Engineering*, 196(17-20):2252–2268.

Cottereau, R., Clouteau, D., and Soize, C. (2008) Probabilistic impedance of foundation, impact of the seismic design on uncertain soils *Earthquake Engineering and Structural Dynamics*, 37(6):899–918.

Das, S. and Ghanem, R. (2009) A bounded random matrix approach for stochastic upscaling *Multiscale Model. Simul.*, 8(1):296325.

Das, S., Ghanem, R., and Finette, S. (2009) Polynomial chaos representation of spatio-temporal random field from experimental measurements *Journal of Computational Physics*, 228:8726–8751.

Das, S., Ghanem, R., and Spall, J. C. (2008) Asymptotic sampling distribution for polynomial chaos representation from data: a maximum entropy and fisher information approach *SIAM Journal on Scientific Computing*, 30(5):2207–2234.

Deb, M., Babuska, I., and Oden, J. (2001) Solution of stochastic partial differential equations using galerkin finite element techniques *Computer Methods in Applied Mechanics and Engineering*, 190:6359–6372.

Debusschere, B., Najm, H., Pebay, P., and *et al.* (2004) Numerical challenges in the use of polynomial chaos representations for stochastic processes *SIAM Journal on Scientific Computing*, 26(2):698–719.

Deodatis, G. and Spanos, P. D. (2008) 5th international conference on computational stochastic mechanics *Special issue of the Probabilistic Engineering Mechanics*, 23(2-3):103–346.

Desceliers, C., Ghanem, R., and Soize, C. (2006) Maximum likelihood estimation of stochastic chaos representations from experimental data *International Journal for Numerical Methods in Engineering*, 66(6):978–1001.

Desceliers, C., Soize, C., and Cambier, S. (2004) Non-parametric - parametric model for random uncertainties in nonlinear structural dynamics - application to earthquake engineering *Earthquake Engineering and Structural Dynamics*, 33(3):315–327.

Desceliers, C., Soize, C., and Ghanem, R. (2007) Identification of chaos representations of elastic properties of random media using experimental vibration tests *Computational Mechanics*, 39(6):831–838.

Desceliers, C., Soize, C., Grimal, Q., Talmant, M., and Naili, S. (2009) Determination of the random anisotropic elasticity layer using transient wave propagation in a fluid-solid multilayer: Model and experiments *Journal of the Acoustical Society of America*, 125(4):2027–2034.

Doostan, A., Ghanem, R., and Red-Horse, J. (2007) Stochastic model reductions for chaos representations *Computer Methods in Applied Mechanics and Engineering*, 196(37-40):3951–3966.

Doostan, A. and Iaccarino, G. (2009) A least-squares approximation of partial differential equations with highdimensional random inputs *Journal of Computational Physics*, 228(12):4332–4345.

Duchereau, J. and Soize, C. (2006) Transient dynamics in structures with nonhomogeneous uncertainties induced by complex joints *Mechanical Systems and Signal Processing*, 20(4):854–867.

Durand, J.-F., Soize, C., and Gagliardini, L. (2008) Structural-acoustic modeling of automotive vehicles in presence of uncertainties and experimental identification and validation *Journal of the Acoustical Society of America*, 124(3):1513–1525.

Faverjon, B. and Ghanem, R. (2006) Stochastic inversion in acoustic scattering *Journal of the Acoustical Society of America*, 119(6):3577–3588.

Fernandez, C., Soize, C., and Gagliardini, L. (2009) Fuzzy structure theory modeling of sound-insulation layers in complex vibroacoustic uncertain sytems - theory and experimental validation *Journal of the Acoustical Society of America*, 125(1):138–153.

Fernandez, C., Soize, C., and Gagliardini, L. (2010) Sound-insulation layer modelling in car computational vibroacoustics in the medium-frequency range *Acta Acustica united with Acustica (AAUWA)*, 96(3):437–444.

Fishman, G. (1996) *Monte Carlo: Concepts, algorithms, and applications* Springer-Verlag, New York.

Fougeaud, C. and Fuchs, A. (1967) *Statistique* Dunod, Paris.

Frauenfelder, P., Schwab, C., and Todor, R. (2005) Finite elements for elliptic problems with stochastic coefficients *Computer Methods in Applied Mechanics and Engineering*, 194(2-5):205–228.

Ganapathysubramanian, B. and Zabaras, N. (2007) Sparse grid collocation schemes for stochastic natural convection problems *Journal of Computational Physics*, 225(1):652–685.

Geman, S. and Geman, D. (1984) Stochastic relaxation, gibbs distribution and the bayesian distribution of images *IEEE Transactions on Pattern Analysis and Machine Intelligence*, Vol PAM I-6(6):721–741.

Ghanem, R. (1999) Ingredients for a general purpose stochastic finite elements formulation *Computer Methods in Applied Mechanics and Engineering*, 168(1-4):19–34.

Ghanem, R. and Dham, S. (1998) Stochastic finite element analysis for multiphase flow in heterogeneous porous media *Transp. Porous Media*, 32:239–262.

Ghanem, R. and Doostan, R. (2006) Characterization of stochastic system parameters from experimental data: A bayesian inference approach *Journal of Computational Physics*, 217(1):63–81.

Ghanem, R., Doostan, R., and Red-Horse, J. (2008) A probability construction of model validation *Computer Methods in Applied Mechanics and Engineering*, 197(29-32):2585–2595.

Ghanem, R. and Ghosh, D. (2007) Efficient characterization of the random eigenvalue problem in a polynomial chaos decomposition *International Journal for Numerical Methods in Engineering*, 72(4):486–504.

Ghanem, R. and Kruger, R. M. (1996) Numerical solution of spectral stochastic finite element systems *Computer Methods in Applied Mechanics and Engineering*, 129:289–303.

Ghanem, R., Masri, S., Pellissetti, M., and Wolfe, R. (2005) Identification and prediction of stochastic dynamical systems in a polynomial chaos basis *Computer Methods in Applied Mechanics and Engineering*, 194(12-16):1641–1654.

Ghanem, R. and Pellissetti, M. (2002) Adaptive data refinement in the spectral stochastic finite element method *Comm. Numer. Methods Engrg.*, 18:141–151.

Ghanem, R. and Red-Horse, J. (1999) Propagation of probabilistic uncertainty in complex physical systems using a stochastic finite element approach *Physica D*, 133(1-4):137–144.

Ghanem, R. and Sarkar, A. (2003) Reduced models for the medium-frequency dynamics of stochastic systems *Journal of the Acoustical Society of America*, 113(2):834–846.

Ghanem, R. and Spanos, P. (1990) Polynomial chaos in stochastic finite elements *Journal of Applied Mechanics - Transactions of the ASME*, 57(1):197–202.

Ghanem, R. and Spanos, P. (2003) *Stochastic Finite Elements: A spectral Approach* (revised edition) Dover Publications, New York.

Ghanem, R. and Spanos, P. D. (1991) *Stochastic finite elements: a spectral approach* Springer-Verlag, New York.

Ghosh, D. and Ghanem, R. (2008) Stochastic convergence acceleration through basis enrichment of polynomial chaos expansions *International Journal for Numerical Methods in Engineering*, 73(2):162–184.

Goller, B., Pradlwarter, H., and Schueller, G. (2009) Robust model updating with insufficient data *Computer Methods in Applied Mechanics and Engineering*, 198(37-40):3096–3104.

Guilleminot, J., Noshadravanb, A., Soize, C., and Ghanem, R. (2011) A probabilistic model for bounded elasticity tensor random fields with application to polycrystalline microstructures *Computer Methods in Applied Mechanics and Engineering*, 200(17-20):1637–1648.

Guilleminot, J. and Soize, C. (2010) A stochastic model for elasticity tensors with uncertain material symmetries *International Journal of Solids and Structures*, 47(22-23):3121–3130.

Guilleminot, J. and Soize, C. (2011) Non-gaussian positive-definite matrix-valued random fields with constrained eigenvalues: application to random elasticity tensors with uncertain material symmetries *International Journal for Numerical Methods in Engineering*, page (To appear).

Guilleminot, J., Soize, C., and Kondo, D. (2009) Mesoscale probabilistic models for the elasticity tensor of fiber reinforced composites: experimental identification and numerical aspects *Mechanics of Materials*, 41(12):1309–1322.

Guilleminot, J., Soize, C., Kondo, D., and Benetruy, C. (2008) Theoretical framework and experimental procedure for modelling volume fraction stochastic fluctuations in fiber reinforced composites *International Journal of Solid and Structures*, 45(21):5567–5583.

Hastings, W. K. (1970) Monte carlo sampling methods using markov chains and their applications *Biometrika*, 109:57–97.

Holmes, P., Lumley, J., and Berkooz, G. (1997) *Turbulence, Coherent Structures, Dynamical Systems and Symmetry* Cambridge University Press, Cambridge.

Jaynes, E. T. (1957) Information theory and statistical mechanics *Physical Review*, 108(2):171–190.

Kaipio, J. and Somersalo, E. (2005) *Statistical and Computational Inverse Problems* Springer-Verlag, New York.

Kassem, M., Soize, C., and Gagliardini, L. (2009) Energy density field approach for low- and medium-frequency vibroacoustic analysis of complex structures using a stochastic computational model *Journal of Sound and Vibration*, 323(3-5):849–863.

Kassem, M., Soize, C., and Gagliardini, L. (2011) Structural partitioning of complex structures in the medium-frequency range. an application to an automotive vehicle *Journal of Sound and Vibration*, 330(5):937–946.

Katafygiotis, L. and Beck, J. (1998) Updating models and their uncertainties. ii: Model identifiability *Journal of Engineering Mechanics - ASCE*, 124(4):463–467.

Kim, K., Mignolet, M. P., and Soize, C. (2007) Stochastic reduced order models for uncertain cantilevered plates in large deformations In *2007 ASME Design Engineering Technical Conferences (IDETC)*, Las Vegas, Nevada, USA.

Knio, O. and LeMaitre, O. (2006) Uncertainty propagation in cfd using polynomial chaos decomposition *Fluid Dynamics Research*, 38(9):616–640.

Kunisch, E. and Volkwein, S. (2001) Galerkin proper orthogonal decomposition methods for parabolic problems *Numerische Mathematik*, 90(1):117–148.

Kunisch, E. and Volkwein, S. (2002) Galerkin proper orthogonal decomposition methods for a general equation in fluid dynamics *SIAM Journal of Numerical Analysis*, 40(1):233–253.

Lawson, C. and Hanson, R. (1974) *Solving Least Squares Problems* Prentice-Hall.

Leissing, T., Soize, C., Jean, P., and Defrance, J. (2010) Computational model for long-range non-linear propagation over urban cities *Acta Acustica united with Acustica (AAUWA)*, 96(5):884–898.

LeMaitre, O. and Knio, O. (2010) *Spectral Methods for Uncerainty Quantification with Applications to Computational Fluid Dynamics* Springer, Heidelberg.

LeMaitre, O. P., Knio, O. M., Najm, H. N., and Ghanem, R. (2002) A stochastic projection method for fluid flow. ii. random process *Journal of Computational Physics*, 181:9–44.

LeMaitre, O. P., Knio, O. M., Najm, H. N., and Ghanem, R. (2004a) Uncertainty propagation using wiener-haar expansions *Journal of Computational Physics*, 197(1):28–57.

LeMaitre, O. P., Najm, H. N., Ghanem, R., and Knio, O. (2004b) Multi-resolution analysis of wiener-type uncertainty propagation schemes *Journal of Computational Physics*, 197(2):502–531.

LeMaitre, O. P., Najm, H. N., Pebay, P. P., Ghanem, R., and Knio, O. (2007) Multi-resolution-analysis scheme for uncertainty quantification in chemical systems *SIAM Journal on Scientific Computing*, 29(2):864–889.

Lucor, D., Meyers, J., and Sagaut, P. (2007) Sensitivity analysis of large-eddy simulations to subgrid-scale-model parametric uncertainty using polynomial chaos *Journal of Fluid Mechanics*, 585:255–279.

Lucor, D., Su, C., and Karniadakis, G. (2004) Generalized polynomial chaos and random oscillators *International Journal for Numerical Methods in Engineering*, 60(3):571–596.

Ma, X. and Zabaras, N. (2009) An efficient bayesian inference approach to inverse problems based on an adaptive sparse grid collocation method *Inverse Problems*, 25(3):Article Number: 035013.

Mace, R., Worden, W., and Manson, G. (2005) Uncertainty in structural dynamics *Special issue of the Journal of Sound and Vibration*, 288(3):431–790.

MacKeown, P. K. (1997) *Stochastic Simulation in Physics* Springer-Verlag, Singapore.

Marzouk, Y. and Najm, H. (2009) Dimensionality reduction and polynomial chaos acceleration of bayesian inference in inverse problems *Journal of Computational Physics*, 228(6):1862–1902.

Marzouk, Y., Najm, H., and Rahn, L. (2007) Stochastic spectral methods for efficient bayesian solution of inverse problems *Journal of Computational Physics*, 224(2):560–586.

Mathelin, L. and LeMaitre, O. (2007) Dual based a posteriori estimation for stochastic finite element method *Comm. App. Math. Comp. Sci.*, 2(1):83–115.

Matthies, H. and Keese, A. (2005) Galerkin methods for linear and nonlinear elliptic stochastic partial differential equations *Computer Methods in Applied Mechanics and Engineering*, 194(12-16):1295–1331.

Matthies, H. G. (2008) Stochastic finite elements: Computational approaches to stochastic partial differential equations *Zamm-Zeitschrift Fur Angewandte Mathematik Und Mechanik*, 88(11):849–873.

Mehta, M. L. (1991) *Random Matrices, Revised and Enlarged Second Edition* Academic Press, New York.

Metropolis, N., Rosenbluth, A., Rosenbluth, M., Teller, A., and Teller, E. (1953) Equations of state calculations by fast computing machines *The Journal of Chemical Physics*, 21(6):1087–1092.

Metropolis, N. and Ulam, S. (1949) The monte carlo method *Journal of American Statistical Association*, 49:335–341.

Mignolet, M. P. and Soize, C. (2007) Stochastic reduced order models for uncertain nonlinear dynamical systems In *IMAC XXV*, Orlando, Florida, USA.

Mignolet, M. P. and Soize, C. (2008a) Nonparametric stochastic modeling of linear systems with prescribed variance of several natural frequencies *Probabilistic Engineering Mechanics*, 23(2-3):267–278.

Mignolet, M. P. and Soize, C. (2008b) Stochastic reduced order models for uncertain nonlinear dynamical systems *Computer Methods in Applied Mechanics and Engineering*, 197(45-48):3951–3963.

Najm, H. (2009) Uncertainty quantification and polynomial chaos techniques in computational fluid dynamics *Journal Review of Fluid Mechanics*, pages 35–52.

Nouy, A. (2007) A generalized spectral decomposition technique to solve a class of linear stochastic partial differential equations *Computer Methods in Applied Mechanics and Engineering*, 196(45-48):4521–4537.

Nouy, A. (2008) Generalized spectral decomposition method for solving stochastic finite element equations: Invariant subspace problem and dedicated algorithms *Computer Methods in Applied Mechanics and Engineering*, 197(51-52):4718–4736.

Nouy, A. (2009) Recent developments in spectral stochastic methods for the numerical solution of stochastic partial differential equations *Archives of Computational Methods in Engineering*, 16(3):251–285.

Nouy, A. (2010) Proper generalized decomposition and separated representations for the numerical solution of high dimensional stochastic problems *Archives of Computational Methods in Engineering*, 17(4):403–434.

Nouy, A. and LeMaitre, O. P. (2009) Generalized spectral decomposition for stochastic nonlinear problems *Journal of Computational Physics*, 228(1):202–235.

Ohayon, R. and Soize, C. (1998) *Structural Acoustics and Vibration* Academic Press, San Diego, London.

Papadimitriou, C., Beck, J., and Au, S. (2000) Entropy-based optimal sensor location for structural model updating *Journal of Vibration and Control*, 6(5):781–800.

Papadimitriou, C., Beck, J., and Katafygiotis, L. (2001) Updating robust reliability using structural test data *Probabilistic Engineering Mechanics*, 16(2):103–113.

Papadrakakis, M. and Kotsopulos, A. (1999) Parallel solution methods for stochastic finite element analysis using monte carlo simulation *Computer Methods in Applied Mechanics and Engineering*, 168(1-4):305–320.

Papadrakakis, M. and Lagaros, N. (2002) Reliability-based structural optimization using neural networks and monte carlo simulation *Computer Methods in Applied Mechanics and Engineering*, 191(32):3491–3507.

Papadrakakis, M. and Papadopoulos, V. (1996) Robust and efficient methods for stochastic finite element analysis using monte carlo simulation *Computer Methods in Applied Mechanics and Engineering*, 134(134):325–340.

Pellissetti, M., Capiez-Lernout, E., Pradlwarter, H., Soize, C., and Schueller, G. I. (2008) Reliability analysis of a satellite structure with a parametric and a non-parametric probabilistic model *Computer Methods in Applied Mechanics and Engineering*, 198(2):344–357.

Peters, B. and Roeck, G. D. (2001) Stochastic system identification for operational modal analysis: A review *Journal of Dynamic Systems Measurement and Control-Transactions of The Asme*, 123(4):659–667.

Pradlwarter, H. and Schueller, G. (1997) On advanced monte carlo simulation procedures in stochastic structural dynamics *International Journal of Non-Linear Mechanics*, 32(4):735–744.

Pradlwarter, H. and Schueller, G. (2010) Local domain monte carlo simulation *Structural Safety*, 32(5):275–280.

Pradlwarter, H. J., Schueller, G. I., and Szekely, G. S. (2002) Random eigenvalue problems for large systems *Computer and Structures*, 80:2415–2424.

Red-Horse, J. and Benjamin, A. (2004) A probabilistic approach to uncertainty quantification with limited information *Reliability Engineering and System Safety*, 85:183–190.

Ritto, T., Soize, C., and Sampaio, R. (2009) Nonlinear dynamics of a drill-string with uncertainty model of the bit-rock interaction *International Journal of Non-Linear Mechanics*, 44(8):865–876.

Ritto, T., Soize, C., and Sampaio, R. (2010) Robust optimization of the rate of penetration of a drill-string using a stochastic nonlinear dynamical model *Computational Mechanics*, 45(5):415–427.

Rubinstein, R. Y. and Kroese, D. P. (2008) *Simulation and the Monte Carlo Method* Second Edition, John Wiley & Sons, New York.

Rupert, C. and Miller, C. (2007) An analysis of polynomial chaos approximations for modeling single-fluid-phase flow in porous medium systems *Journal of Computational Physics*, 226(2):2175–2205.

Sakamoto, S. and Ghanem, R. (2002) Polynomial chaos decomposition for the simulation of non-gaussian nonstationary stochastic processes *Journal of Engineering Mechanics-ASCE*, 128(2):190–201.

Sampaio, R. and Soize, C. (2007a) On measures of non-linearity effects for uncertain dynamical systems - application to a vibro-impact system *Journal of Sound and Vibration*, 303(3-5):659–674.

Sampaio, R. and Soize, C. (2007b) Remarks on the efficiency of pod for model reduction in nonlinear dynamics of continuous elastic systems *International Journal for Numerical Methods in Engineering*, 72(1):22–45.

Schueller, G. (2009) Efficient monte carlo simulation procedures in structural uncertainty and reliability analysis - recent advances *Structural Engineering and Mechanics*, 32(1):1–20.

Schueller, G. and Pradlwarter, H. (2009a) Uncertain linear systems in dynamics: Retrospective and recent developments by stochastic approaches *Engineering Structures*, 31(11):2507–2517.

Schueller, G. and Pradlwarter, H. (2009b) Uncertainty analysis of complex structural systems *International Journal for Numerical Methods in Engineering*, 80(6-7):881–913.

Schueller, G. I. (2005a) Computational methods in stochastic mechanics and reliability analysis *Special issue of Computer Methods in Applied Mechanics and Engineering*, 194(12-16):1251–1795.

Schueller, G. I. (2005b) Uncertainties in structural mechanics and analysis-computational methods *Special issue of Computer and Structures*, 83(14):1031–1150.

Schueller, G. I. (2006) Developments in stochastic structural mechanics *Archive of Applied Mechanics*, 75(10-12):755–773.

Schueller, G. I. (2007) On the treatment of uncertainties in structural mechanics and analysis *Computer and Structures*, 85(5-6):235–243.

Schueller, G. I. and Jensen, H. A. (2008) Computational methods in optimization considering uncertainties - an overview *Computer Methods in Applied Mechanics and Engineering*, 198(1):2–13.

Serfling, R. J. (1980) *Approximation Theorems of Mathematical Statistics* John Wiley & Sons.

Shannon, C. E. (1948) A mathematical theory of communication *Bell System Technology Journal*, 27(14):379–423 & 623–659.

Soize, C. (1980) Oscillators submitted to squared gaussian processes *Journal of Mathematical Physics*, 21(10):2500–2507.

Soize, C. (1994) *The Fokker-Planck Equation for Stochastic Dynamical Systems and its Explicit Steady State Solutions* World Scientific Publishing Co Pte Ltd, Singapore.

Soize, C. (2000) A nonparametric model of random uncertainties on reduced matrix model in structural dynamics *Probabilistic Engineering Mechanics*, 15(3):277–294.

Soize, C. (2001) Maximum entropy approach for modeling random uncertainties in transient elastodynamics *Journal of the Acoustical Society of America*, 109(5):1979–1996.

Soize, C. (2003a) Random matrix theory and non-parametric model of random uncertainties *Journal of Sound and Vibration*, 263(4):893–916.

Soize, C. (2003b) Uncertain dynamical systems in the medium-frequency range *Journal of Engineering Mechanics*, 129(9):1017–1027.

Soize, C. (2005a) A comprehensive overview of a non-parametric probabilistic approach of model uncertainties for predictive models in structural dynamics *Journal of Sound and Vibration*, 288(3):623–652.

Soize, C. (2005b) Random matrix theory for modeling uncertainties in computational mechanics *Computer Methods in Applied Mechanics and Engineering*, 194(12-16):1333–1366.

Soize, C. (2006) Non gaussian positive-definite matrix-valued random fields for elliptic stochastic partial differential operators *Computer Methods in Applied Mechanics and Engineering*, 195(1-3):26–64.

Soize, C. (2008a) Construction of probability distributions in high dimension using the maximum entropy principle. applications to stochastic processes, random fields and random matrices *International Journal for Numerical Methods in Engineering*, 76(10):1583–1611.

Soize, C. (2008b) Tensor-valued random fields for meso-scale stochastic model of anisotropic elastic microstructure and probabilistic analysis of representative volume element size *Probabilistic Engineering Mechanics*, 23(2-3):307–323.

Soize, C. (2010a) Generalized probabilistic approach of uncertainties in computational dynamics using random matrices and polynomial chaos decompositions *International Journal for Numerical Methods in Engineering*, 81(8):939–970.

Soize, C. (2010b) Identification of high-dimension polynomial chaos expansions with random coefficients for non-gaussian tensor-valued random fields using partial and limited experimental data *Computer Methods in Applied Mechanics and Engineering*, 199(33-36):2150–2164.

Soize, C. (2010c) Random matrices in structural acoustics In Wright, M. and Weaver, R., editors, *New Directions in Linear Acoustics: Random Matrix Theory, Quantum Chaos and Complexity*, pages 206–230. Cambridge University Press, Cambridge.

Soize, C. (2011) A computational inverse method for identification of non-gaussian random fields using the bayesian approach in very high dimension *Computer Methods in Applied Mechanics and Engineering*, Accepted for publication, on line July 15, 2011, doi: 10.1016/j.cma.2011.07.005.

Soize, C., Capiez-Lernout, E., Durand, J.-F., Fernandez, C., and Gagliardini, L. (2008a) Probabilistic model identification of uncertainties in computational models for dynamical systems and experimental validation *Computer Methods in Applied Mechanics and Engineering*, 198(1):150–163.

Soize, C., Capiez-Lernout, E., and Ohayon, R. (2008b) Robust updating of uncertain computational models using experimental modal analysis *AIAA Journal*, 46(11):2955–2965.

Soize, C. and Chebli, H. (2003) Random uncertainties model in dynamic substructuring using a nonparametric probabilistic model *Journal of Engineering Mechanics*, 129(4):449–457.

Soize, C. and Desceliers, C. (2010) Computational aspects for constructing realizations of polynomial chaos in high dimension *SIAM Journal On Scientific Computing*, 32(5):2820–2831.

Soize, C. and Ghanem, R. (2004) Physical systems with random uncertainties : Chaos representation with arbitrary probability measure *SIAM Journal On Scientific Computing*, 26(2):395–410.

Soize, C. and Ghanem, R. (2009) Reduced chaos decomposition with random coefficients of vector-valued random variables and random fields *Computer Methods in Applied Mechanics and Engineering*, 198(21-26):1926–1934.

Spall, J. C. (2003) *Introduction to Stochastic Search and Optimization* John Wiley.

Stefanou, G., Nouy, A., and Clément, A. (2009) Identification of random shapes from images through polynomial chaos expansion of random level set functions *International Journal for Numerical Methods in Engineering*, 79(2):127–155.

Szekely, G. and Schuller, G. (2001) Computational procedure for a fast calculation of eigenvectors and eigenvalues of structures with random properties *Computer Methods in Applied Mechanics and Engineering*, 191(8-10):799–816.

Ta, Q., Clouteau, D., and Cottereau, R. (2010) Modeling of random anisotropic elastic media and impact on wave propagation *European Journal of Computational Mechanics*, 19(1-2-3):241253.

Taflanidis, A. and Beck, J. (2008) An efficient framework for optimal robust stochastic system design using stochastic simulation *Computer Methods in Applied Mechanics and Engineering*, 198(1):88–101.

Tan, M. T., Tian, G.-L., and Ng, K. W. (2010) *Bayesian Missing Data Problems, EM, Data Augmentation and Noniterative Computation* Chapman & Hall / CRC Press, Boca Raton.

Terrell, G. R. and Scott, D. W. (1992) Variable kernel density estimation *The Annals of Statistics*, 20(3):1236–1265.

Walter, E. and Pronzato, L. (1997) *Identification of Parametric Models from Experimental Data* Springer-Verlag, Berlin.

Wan, X. and Karniadakis, G. (2005) An adaptive multi-element generalized polynomial chaos method for stochastic differential equations *Journal of Computational Physics*, 209(2):617–642.

Wan, X. and Karniadakis, G. (2006) Multi-element generalized polynomial chaos for arbitrary probability measures *SIAM Journal on Scientific Computing*, 28(3):901–928.

Wan, X. and Karniadakis, G. (2009) Error control in multielement generalized polynomial chaos method for elliptic problems with random coefficients *Comm. Comput. Phys.*, 5(2-4):793–820.

Wang, J. and Zabaras, N. (2004) A bayesian inference approach to the inverse heat conduction problem *International Journal of Heat and Mass Transfer*, 47(17-18):3927–3941.

Wang, J. and Zabaras, N. (2005) Hierarchical bayesian models for inverse problems in heat conduction *Inverse Problems*, 21(1):183–206.

Wang, X., Mignolet, M., Soize, C., and Khannav, V. (2010) Stochastic reduced order models for uncertain infinite-dimensional geometrically nonlinear dynamical system - stochastic excitation cases In *IUTAM Symposium on Nonlinear Stochastic Dynamics and Control*, Hangzhou, China.

Webster, C., Nobile, F., and Tempone, R. (2007) A sparse grid stochastic collocation method for partial differential equations with random input data *SIAM Journal on Numerical Analysis*, 46(5):2309–2345.

Wiener, N. (1938) The homogeneous chaos *American Journal of Mathematics*, 60(1):897–936.

Wright, M. and Weaver, R. (2010) *New Directions in Linear Acoustics: Random Matrix Theory, Quantum Chaos and Complexity* Cambridge University Press, Cambridge.

Xiu, D. and Karniadakis, G. (2002a) Modeling uncertainty in steady state diffusion problems via generalized polynomial chaos *Computer Methods in Applied Mechanics and Engineering*, 191(43):4927–4948.

Xiu, D. and Karniadakis, G. (2002b) Wiener-askey polynomial chaos for stochastic differential equations *SIAM Journal on Scientific Computing*, 24(2):619–644.

Xiu, D. and Karniadakis, G. (2003) Modeling uncertainty in flow simulations via generalized polynomial chaos *Journal of Computational Physics*, 187(1):137–167.

Zabaras, N. and Ganapathysubramanian, B. (2008) A scalable framework for the solution of stochastic inverse problems using a sparse grid collocation approach *Journal of Computational Physics*, 227(9):4697–4735.

Index

D

E

S

T

U